T0329090

HIGH-ORDER MODELS IN SEMANTIC IMAGE SEGMENTATION

HIGH-ORDER MODELS IN SEMANTIC IMAGE SEGMENTATION

ISMAIL BEN AYED
Systems Engineering Department
ETS Montreal
Montreal, QC, Canada

ACADEMIC PRESS
An imprint of Elsevier

Academic Press is an imprint of Elsevier
125 London Wall, London EC2Y 5AS, United Kingdom
525 B Street, Suite 1650, San Diego, CA 92101, United States
50 Hampshire Street, 5th Floor, Cambridge, MA 02139, United States
The Boulevard, Langford Lane, Kidlington, Oxford OX5 1GB, United Kingdom

Notices

Knowledge and best practice in this field are constantly changing. As new research and experience
broaden our understanding, changes in research methods, professional practices, or medical
treatment may become necessary.

Practitioners and researchers must always rely on their own experience and knowledge in
evaluating and using any information, methods, compounds, or experiments described herein. In
using such information or methods they should be mindful of their own safety and the safety of
others, including parties for whom they have a professional responsibility.

To the fullest extent of the law, neither the Publisher nor the authors, contributors, or editors,
assume any liability for any injury and/or damage to persons or property as a matter of products
liability, negligence or otherwise, or from any use or operation of any methods, products,
instructions, or ideas contained in the material herein.

ISBN: 978-0-12-805320-1

For information on all Academic Press publications
visit our website at https://www.elsevier.com/books-and-journals

Publisher: Mara Conner
Acquisitions Editor: Tim Pitts
Editorial Project Manager: Naomi Robertson
Production Project Manager: Prasanna Kalyanaraman
Cover Designer: Matthew Limbert

Typeset by VTeX

Working together
to grow libraries in
developing countries

www.elsevier.com • www.bookaid.org

Contents

General introduction

General context

Automatically recognizing semantic concepts in numeric images, e.g., a specific organ or disease in a medical image, or a specific target object in a video sequence, is a computer vision research subject of paramount importance, due to its theoretical and methodological challenges as well as breadth of applications. Scene-understanding algorithms would enable machines to extract very valuable information from visual data, with an enormous potential to enable powerful predictive models, to provide new insights and to guide decisions. For instance, semantic segmentation, which is the main focus of this book, is of pivotal importance in various image interpretation systems. This difficult and cumbersome computational task aims at assigning automatically a semantic label (or class) to every pixel in an image, e.g., a "car" or a "person" in a street scene, acquired by a standard camera or a tumor structure in a medical MRI. In medical imaging, segmentation information translates directly into quantitative measures for disease diagnosis, treatment and follow-up, surgery planning, radiology reporting, and health-care research and practices at large. Powerful and accurate segmentation algorithms have the potential to disrupt extremely important application areas of medical imaging, such as cardiology, neuroscience and oncology. More generally, the problem is of pivotal importance in a breadth of application areas, such as autonomous driving, surveillance, remote sensing, industrial inspections, and web-based commercial platforms, just to name a few examples.

From an algorithmic perspective, visual recognition problems, including semantic segmentation, are very challenging and domain-dependent because of the diversity, variability and complexity of real-world scenes. The current overwhelming growth of large-scale image data acquired and stored everyday, coupled with the recent computational and algorithmic breakthroughs in deep learning research, have yielded unprecedented advances in visual-recognition performances in the last decade, triggering wide interest in both academia and industry. There is a clear consensus that, in the next decades, such algorithms will disrupt strategic areas such as health care, security, autonomous systems and robotics, remote sensing, manufacturing, and social media, among many others. However, this also raises substantial computational and algorithmic challenges, which call for bringing current image interpretation algorithms to a whole new level of automation, scalability and accuracy.

From graphical models to deep learning

Semantic segmentation has been traditionally tackled using graphical models known as a Conditional Random Fields (CRFs). CRFs explicitly model the relationships between the latent pixel predictions, typically embedding some prior knowledge about the sought segmentation regions. The most basic form of CRFs encodes some relationships between pairs of pixels, such as the classical Potts model, which encourages segmentation boundary smoothness and alignment with the image edges. Beyond basic pairwise interactions, higher-order interactions between the variables being estimated enable the embeding of very useful prior information on the solutions. Such priors take the form of *high-order functions*, which have been intensively investigated in the CRF literature. Depending on the application, priors could come from different sources of side information that might be associated with the images. For instance, in medical imaging, it is common to have access to some prior knowledge about the size or shape of the target segmentation regions. In oncology, for example, approximate tumor size information might be available in the radiology text reports associated with the images. Such prior knowledge does not have to be very precise. Yet, imposing such priors on segmentation objectives, in the form of high-order functions, could help in mitigating the difficulty of the task. Beyond segmentation, high-order functions occur in a breadth of problems in machine learning, such as unsupervised data clustering.

This book discusses various high-order functions in image segmentation and, more generally, in the broad context of unsupervised data clustering. The treatment we provide in several chapters focuses on discrete binary assignment variables, including pairwise and high-order objective functions that originated from the rich and classical literature on CRFs and clustering. We also discuss how inference with respect to the discrete hidden variables could benefit from powerful and scalable combinatorial optimization techniques. Several among these discrete-variable objectives, initially studied in CRFs or in data clustering, have recently motivated interesting loss functions in the context of deep learning models, either during the training or inference phases. Such unsupervised loss functions leverage unlabeled data by embedding useful priors. In this setting, the probability outputs of a deep network could be viewed as soft, parameterized versions of the discrete binary indicator functions of the segmentation regions, or of the clusters in the context of data clustering. We have dedicated several chapters that discuss examples of unsupervised high-order and pairwise CRF loss functions defined over the outputs of a deep network. In these cases, optimization could be carried out with gradient descent, the workhorse for training deep networks. We also discuss hybrid discrete-continuous solutions, which could benefit from powerful discrete optimization techniques in the context of training deep networks.

A large body of work in the recent computer vision and machine learning literature points to a consensus: leveraging unlabeled imaging data with some side prior knowledge that might be associated with the images (e.g., text) is fundamental, and has great potential to mitigate the lack of labels and improve the generalization of deep learning models. Indeed, in cases where the training data is representative enough, i.e., the images at the target conditions have characteristics that are quite similar to the training data, standard supervised deep-learning models can achieve excellent performance. However, the generalization of such models is seriously challenged when the target-condition samples differ from the training data. This occurs often in practice, for instance, when dealing with new classes, either unseen or rarely represented in the training data, or when dealing with data shifts, i.e., when the images in the target conditions differ from those seen in training. For instance, in medical imaging, such a major data-shift challenge is very common due the changes in imaging protocols, clinical sites and subject populations.

In a breadth of applications, current supervised deep learning models are typically trained and evaluated on limited numbers of images and semantic classes. Therefore, despite their huge recent impact, they may have difficulty capturing the substantial variability encountered in real scenarios. Building manually labeled data covering the overwhelming amounts of real-world scenarios is infeasible because this would require unreasonably huge efforts by humans. In particular, this is the case when supervision requires scarce human-expert knowledge, as is the case in medical imaging, or when dealing with prohibitively time-consuming and dense-prediction tasks involving the manual labeling of millions of pixels per image, as in semantic segmentation. For instance, the most comprehensive and publicly available semantic color image-segmentation benchmark has fewer than 30 classes, and focuses specifically on street scenes. In medical image segmentation, the labels are typically restricted to a very few anatomical structures and small data sets, a difficulty further compounded by the complexity of the data, e.g., 3D, multi-modal or temporal data. The unsupervised learning concepts discussed in this book, which stem from the classical literature in graphical models and clustering, should provide a good technical background to inspire deep learning researchers in the design of unsupervised learning models enhanced with useful and domain-specific prior knowledge. Also, the optimization techniques discussed in this book could inspire interesting hybrid discrete-continuous solutions for training deep networks that go beyond basic gradient descent. Moreover, the interactive (user-guided) segmentation methods discussed in this book could help practitioners design fast labeling tools for building large labeled data sets.

Chapter 1

Chapter 1 introduces the basic concepts in Conditional Random Fields (CRFs), which are very popular in computer vision. CRFs are graphical models. They describe images as graphs, whose nodes are associated to hidden variables, for instance, the latent label assignments representing a segmentation of a given image. Such graph representations are very convenient because they connect explicitly the hidden variables. The chapter focuses on the most basic form of connections that involve pairs of nodes and can be viewed as weighted edges on the graph. Such pairwise connections embed statistical dependence between the hidden variables, typically enforcing some prior knowledge. We take pairwise regularization priors, such as the widely used Potts model, as examples. Also, we examine these graphical models from Bayesian-estimation and energy-minimization perspectives, and discuss how inference with respect to the hidden variables can benefit from powerful and scalable optimization techniques. While the discussion is limited to discrete variables, the basics provided in this chapter are a prerequisite for understanding and using CRF loss functions, which have been widely used recently in training deep networks. A full chapter (Chapter 9) is dedicated to the use of CRFs in the context of deep learning.

Chapter 2

Chapter 2 discusses combinatorial graph cuts, which have made a substantial impact in computer vision for optimizing effectively an important class of CRF objectives, which is common in image segmentation. These powerful discrete optimizers can provide exact global optima for binary-labeling problems and competitive approximate solutions with optimality bounds for multi-label problems. We discuss the basics of finding the minimum-cut/maximum-flow of an S-T graph for binary-labeling problems, along with standard max-flow algorithms, including the very popular Boykov–Kolmogorov (BK) algorithm. Furthermore, we detail well-known move-making algorithms for tackling multi-label problems. We discuss optimality guarantees, the conditions under which these powerful solvers could be used, as well as computational-efficiency aspects for computer vision problems.

Chapter 3

Chapter 3 discusses variational mean-field inference, which is widely used in computer vision and machine learning. The focus is on how to tackle efficiently CRFs via a well-known parallel message-passing algorithm, which has been useful in a large body of recent semantic segmentation works. After giving a Majorize–Minimize (MM) perspective of this message-passing technique, we examine a generalization that enables parallel and convergent updates for a general form of pairwise CRFs. We discuss the aspects related to computational scalability for computer vision problems with large numbers (millions) of variables as is the case in semantic image segmentation.

Chapter 4

Chapter 4 discusses classical segmentation formulations based on maximum-likelihood model fitting and regularization. Generally, this class of techniques optimizes high-order functions that integrate two types of terms, one data term penalizing the deviation of image features within each region from parametric statistical models and one regularizer evaluating segmentation boundary length. We review very popular algorithms such as GrabCut and the Chan–Vese model, among others. We examine the link between these segmentation formulations and widely used data clustering functions, such as K-means and its probabilistic or kernel-induced generalizations.

Chapter 5

Chapter 5 gives information-theoretic perspectives on several popular segmentation algorithms based on maximum-likelihood model fitting. These algorithms optimize high-order functions, which correspond to maximizing the mutual information between the inputs features and latent labeling. We examine the link between maximum-likelihood estimation and entropy minimization via a generative view of the mutual information. Then, we examine a different, discriminative view of the mutual information, which reveals interesting connections to discriminative clustering models. Such information-based clustering models have been recently used in several subareas in machine learning, including unsupervised representation learning, unsupervised domain adaptation and few-shot learning. Furthermore, the mutual-information insights discussed in this chapter explain some class-balance artifacts that are observed in practice for popular likelihood-based segmentation algorithms such as GrabCut. This prescribes principled methodologies for dealing with such artifacts via optimizing some high-order functions. The discussion is supported by several experimental illustrations that include real and synthetic data.

Chapter 6

This chapter discusses several types of high-order functionals and their usefulness in image segmentation and data clustering. This includes priors on the sizes and shapes of the target segmentation regions, balanced graph-clustering objectives such as the popular normalized cut, and distribution-matching priors. The chapter is supported by several experimental examples.

Chapter 7

High-order functions are useful in image segmentation and data clustering, but may lead to challenging optimization problems. Unfortunately, powerful and global optimization techniques, such as graph cuts, are restricted to special forms of regional and boundary functions and, in general, cannot deal directly with high-order regional terms. Bound and pseudo-bound optimization can address a wide range of high-order functions, and these are also discussed in this chapter.

Chapter 8

This chapter discusses iterative trust-region optimization, which splits a difficult problem into a sequence of easier subproblems, each corresponding to a local approximation of the high-order terms in the objective function.

The approximation is "trusted" only within a region at the vicinity of the current solution. At each iteration, the global optimum of the approximation is evaluated, and the size of the trust region is updated, using some measure of the quality of the current approximation. We discuss how to use the trust-region framework in image segmentation, and detail a specific medical-imaging example based on a high-order term enforcing a shape prior. We also provide details as to how to derive approximations of high-order functions using *Gateâux* derivatives.

Chapter 9

Weakly-supervised learning methods, which use limited sets of annotated images and unlabeled data, are currently attracting wide research interest. Imposing priors on deep networks can mitigate the lack of annotations, leveraging unlabeled samples with prior knowledge. This chapter discusses how some of the pairwise or high-order regularizers discussed in the book could be very useful in weakly supervised segmentation. More specifically, we discuss several loss functions that impose regularization and edge alignment priors on the probability outputs of a deep network. Adding unsupervised loss terms, such as pairwise CRF regularization or high-order balanced graph clustering, can achieve performances close to full supervision, while using only fractions of the ground-truth labels. We discuss various optimization strategies for regularization losses in the context of deep-network segmentation. In particular, we discuss a general alternating direction method strategy for training neural networks that makes it possible to directly take advantage of powerful discrete graph-cut solvers for CRF losses, providing an effective optimization alternative to standard gradient descent.

Chapter 10

This chapter focuses on the setting in which one has some prior knowledge on the target segmentation regions. For instance, in medical imaging, it is common to have some prior knowledge about the size or the shape of a target organ. Such knowledge does not have to be precise, and may take the imprecise form of lower and upper bounds on region properties. We discuss how to complement such priors using inequality constraints on functions of the network outputs, and the usefulness of such constrained formulations in the context of weakly supervised segmentation. Furthermore, we give detailed descriptions of various optimization strategies for these constrained deep models, including Lagrangian–dual optimization and penalty methods for handling inequality constraints.

CHAPTER 1

Markov random fields

1.1 Discrete representations

Discrete random fields, which are often referred to as Markov or conditional random fields (MRFs/CRFs), are a major research subject in computer vision [1] for the methodological convenience they afford in modeling images and the substantial impact they have made over a broad range of realistic computer vision applications, for instance, segmentation, motion estimation, image registration, stereo, object recognition and image editing, among many others. The following list summarizes the main concepts underlying discrete random fields applied to computer vision:

- MRFs/CRFs use graph representations, where an image is viewed as a set of nodes, each corresponding to a pixel or a group of pixels.
- Each node is associated with a *hidden* variable, e.g., a discrete binary array indicating a segmentation of the image into a foreground and a background. Furthermore, nodes or groups of nodes are associated with one or many observed image values, e.g., colors and textures, among others.
- MRFs/CRFs often construct a probabilistic model or, equivalently, an *energy* function, that references both hidden and observed variables. Typically, such models account for two main aspects: (i) conformity of the solution to the observed data and (ii) some prior knowledge that we might have or learn about the problem.
- Inference with respect to the hidden variables uses powerful optimization techniques. One of the main reasons behind the practical success of discrete random fields in vision is the development of several key inference algorithms that can handle the computational burden of real computer vision problems, while providing some optimality guarantees. In the next chapters, we will discuss in further detail some of the most influential algorithms for optimizing random field energies in vision.

Discrete graph-based representations are very convenient in computer vision because the graphs connect the hidden variables explicitly. The most basic form of connections involves pairs of nodes and are described as weighted edges on the graph. Such *pairwise* connections allow embedding statistical dependence between hidden variables, which are encoded in the form of edge weights; examples will follow soon in this chapter. Typically, but not always, such pairwise connections embed some *prior knowledge* that we have about the problem at hand. For instance, spatial coherence priors, which are very popular in computer vision, use the fact that nearby pixels are likely to have the same hidden variables. For instance, take the task of segmenting an image into two regions (or segments), a foreground and a background. In this case, each node (pixel) p is associated with a binary hidden variable $x_p \in \{0, 1\}$, with $x_p = 1$ indicating that p is a foreground pixel and $x_p = 0$ indicating a background pixel. To embed a prior based on the fact that neighboring pixels (p, q) are likely to belong to the same segment, we associate some probabilistic bias to each edge. Such a bias is embedded in the energy function in the form of *pairwise potentials*, a basic example of which we shall see later in this chapter.

In computer vision problems, pairwise edges typically involve only nearby pixels. Connecting all pixels (nodes) will result in fully (densely) connected graphs. As we will discuss in more detail later in the book, applying standard inference methods to such dense graphs will be computationally intractable, and special care has to be made to design efficient inference methods for images of typical sizes [2,3].

1.1.1 Markov random fields (MRFs) on graphs

In many computer vision and image processing tasks, it is common to solve a discrete energy optimization problem. We infer the optimal values of hidden discrete random variables $\mathbf{X} = (X_p)_{p \in \Omega}$, which describe a pixelwise labeling of an observed image feature $\mathbf{f} = (f_p)_{p \in \Omega}$, with $\Omega \subset \mathbb{R}^{2,3}$ the spatial image domain. The possible values of each X_p is a finite set of discrete labels $\{0, \ldots, L-1\}$. In semantic segmentation of color images, for instance, each label describes a semantic category, e.g., "car", "sky", "bicycle", etc. A labeling is a particular assignment $\mathbf{x} = (x_p)_{p \in \Omega}$ to $\mathbf{X}$, which assigns a label $x_p \in \{0, \ldots, L-1\}$ to pixel p.

We associate $\mathbf{X}$ to a Markov random field (MRF), which is the graph [1]

$$\mathcal{G} = \langle \mathcal{V}, \mathcal{W} \rangle, \tag{1.1}$$

where $\mathcal{V}$ is a set of nodes (vertices) of the graph, and $\mathcal{W}$ a set of undirected edges connecting these nodes. Typically, each node p in $\mathcal{V}$ corresponds to a pixel or a superpixel[1] of the image, and the corresponding random variable X_p takes one value x_p within label set $\{0, \ldots, L - 1\}$. For the task of image segmentation, for instance, each label $l \in \{0, \ldots, L - 1\}$ corresponds to a segment (subregion) of image domain Ω,

$$S^l = \{p \in \Omega \mid x_p = l\}. \tag{1.2}$$

Let N_p denotes the neighborhood of node p, i.e., the set of nodes adjacent (connected) to p: $N_p = \{q \in \mathcal{V}, q \neq p \mid (p, q) \in \mathcal{W}\}$. A MRF satisfies the following property for conditional probabilities:

$$\Pr\left(X_p = x_p \mid \{X_q = x_q\}_{q \in \mathcal{V}, q \neq p}\right) = \Pr\left(X_p = x_p \mid \{X_q = x_q\}_{q \in N_p}\right). \tag{1.3}$$

This means that the random variable of a given node p depends only on the random variables within the corresponding neighborhood N_p, often referred to as the Markov blanket of node p in the context of MRFs [1]. Following the *Hammersley–Clifford* theorem, the joint distribution over the MRF variables satisfying (1.3) can be expressed as a product of positive functions, each corresponding to a *maximal clique* c:

$$\Pr(\mathbf{X} = \mathbf{x}) = \prod_{c \in C} F_c(\mathbf{x}_c) \tag{1.4}$$

where:

- C is the set of maximal cliques of graph $\mathcal{G}$. A maximal clique $c \in C$ is a fully connected subgraph of $\mathcal{G}$, i.e., any additional node would violate the full connectedness of the clique.
- Clique factor F_c is a nonnegative function defined over the assignments of the random variables associated with clique c: $\mathbf{x}_c = \{x_p \mid p \in c\}$.

The MRF distribution in (1.4) is commonly written in terms of an energy function $\mathcal{R}(\mathbf{x})$, taking the form of a *Gibbs* distribution [1]:

$$\Pr(\mathbf{X} = \mathbf{x}) = \frac{1}{Z} \exp\left(-\mathcal{R}(\mathbf{x})\right), \text{ with}$$

---

[1] Superpixels are groups of pixels obtained from an unsupervised segmentation algorithm, e.g., [4]. Typically, superpixels are homogeneous segments (in term of image data such as color) whose boundaries adhere to image edges. The use of superpixels can ease significantly the computational burden associated with such graphical models because inference needs to be done over a small number of superpixels rather than the whole image domain [5].

$$\mathcal{R}(\mathbf{x}) = \sum_{c \in C} \psi_c(\mathbf{x}_c) \tag{1.5}$$

Here, *clique potential* $\psi_c(\mathbf{x}_c)$ is a positive function defined over the random variables of clique c, and Z is a partition function enforcing the normalization condition for distributions.

A breadth of computer vision problems, including segmentation, can be stated as a Bayesian estimation, assuming the MRF distribution in (1.5) is a prior and given a set of data observation $\mathbf{f} = (f_p)_{p \in \Omega}$ (e.g., image colors, intensities, responses of different convolutions):

$$\Pr(\mathbf{X} = \mathbf{x}|\mathbf{f}) \propto \Pr(\mathbf{f}|\mathbf{X} = \mathbf{x}) \Pr(\mathbf{X} = \mathbf{x}) \tag{1.6}$$

In this case, the problem is a *maximum a posteriori* (MAP) estimation of the following general form:

$$\hat{\mathbf{x}} = \operatorname{argmax}_{\mathbf{x}} \Pr(\mathbf{X} = \mathbf{x}|\mathbf{f}) = \operatorname{argmax}_{\mathbf{x}} \Pr(\mathbf{f}|\mathbf{X} = \mathbf{x}) \Pr(\mathbf{X} = \mathbf{x}) \tag{1.7}$$

where $\Pr(\mathbf{f}|\mathbf{X} = \mathbf{x})$ is the *likelihood* of observations given the hidden variables of $\mathbf{X}$. A common choice for this MAP estimation in computer vision is to assume that the data observations f_p are independent random variables when conditioned on the MRF hidden variables x_p:

$$\Pr(\mathbf{f}|\mathbf{X} = \mathbf{x}) = \prod_{p \in \Omega} \Pr(f_p|X_p = x_p) \tag{1.8}$$

Taking minus the logarithm of the posterior in (1.6), and putting everything together, we arrive at minimizing an objective function of the following general form:

$$\mathcal{E}(\mathbf{x}) = \sum_{p \in \Omega} \psi_p(x_p) + \sum_{c \in C} \psi_c(\mathbf{x}_c) \tag{1.9}$$

Let us start by examining a simple example of MRF model (1.9) for segmenting an image into two regions, a foreground and a background. In this example, pioneered in the seminal segmentation work of Boykov and Jolly [6], the objective function contains a sum of *unary potentials*, which take the form of pixel log-likelihoods measuring the conformity of image observations within each segment to some fixed, *a priori* estimated probabilistic model, and *pairwise potentials*, whose optimization favors segmentation boundaries that are smooth (or regularized) and aligned with image edges, i.e., strong changes in color.

1.1.2 Unary potentials

In the general MRF form in Eq. (1.9), $\psi_p(l)$ is a unary potential, which measures the penalty that we pay if label $l \in \{0, \ldots, L - 1\}$ is assigned to pixel p. A typical choice for unary potentials is the negative log-likelihoods [6]:

$$\psi_p(l) = -\ln \Pr(\mathbf{f}_p | X_p = l, \boldsymbol{\theta}), \tag{1.10}$$

where $\Pr(\mathbf{f}_p | X_p = l, \boldsymbol{\theta})$ denotes some given generative probability model of image features within segmentation region S^l. $\boldsymbol{\theta}$ is a set of parameters characterizing these probability models (or distributions), and $\mathbf{f}_p \in \mathbb{R}^N$ denotes the observed feature at spatial location p. These features might be intensity values in $\mathbb{R}^1$, colors in $\mathbb{R}^3$ or other features in some higher-dimensional spaces, e.g., the outputs of several convolutions of the image. For instance, for color features, probability model $\Pr(\mathbf{f}_p | X_p = l, \boldsymbol{\theta})$ might take the form of Gaussian mixture models (GMM) or histograms [6–8]. Let us consider the simple case of two-region segmentation, a foreground $S^1 = \{p \in \Omega \,|\, x_p = 1\}$ and a background $S^0 = \{p \in \Omega \,|\, x_p = 1\}$, with probability models $\Pr(\mathbf{f}_p | X_p = l, \boldsymbol{\theta})$, $l \in \{0, 1\}$, known *a priori*. In practice, the models can be learned from simple user inputs,[2] e.g., a few scribbles as illustrated in Fig. 1.2, or from training images. For tutorial purposes, let us assume that we have "perfect" knowledge of the foreground and background probability models, e.g., by estimating these from the ground-truth mask, which is a semantic segmentation performed manually by a user. In the two-region ground-truth depicted in Fig. 1.1, the foreground region corresponds to the semantic category "soldier" and the background to the rest of the image domain. Of course, this assumption is not valid in practice because we do not have access to a "perfect" segmentation ground-truth of the image we intend to segment. However, as it will become clear shortly, this will help us understand the practical value of pairwise MRF potentials for regularizing segmentation boundaries. For this two-region problem, it is easy to see that the summation of unary potentials in Eq. (1.9), expressed as $\psi_p(l) = -\ln \Pr(\mathbf{f}_p | X_p = l, \boldsymbol{\theta})$, $l \in \{0, 1\}$, can be expressed as summations over

---

[2] Such a scenario based on simple user inputs is of high practical value in various applications, for instance, fast image editing. It can also be useful for synthesizing a large amount of masks for training semantic segmentation algorithms [9], assuming that the training images come with weak annotations such as scribbles [10] or bounding boxes [9].

latent segmentation regions S^0 and S^1:

$$\sum_{p \in \Omega} \psi_p(x_p) = -\sum_{p \in S^0} \ln \Pr(\mathbf{f}_p | X_p = 0, \boldsymbol{\theta}) - \sum_{p \in S^1} \ln \Pr(\mathbf{f}_p | X_p = 1, \boldsymbol{\theta}). \qquad (1.11)$$

It is straightforward to write this expression in terms of binary variables $x_p \in \{0, 1\}$, which will make it easy to understand the effect of optimizing unary potentials in the case of two regions:

$$\sum_{p \in \Omega} \psi_p(x_p) = -\sum_{p \in \Omega}(1 - x_p) \ln \Pr(\mathbf{f}_p | X_p = 0, \boldsymbol{\theta}) - \sum_{p \in \Omega} x_p \ln \Pr(\mathbf{f}_p | X_p = 1, \boldsymbol{\theta})$$

$$\stackrel{c}{=} \sum_{p \in \Omega} x_p \ln \frac{\Pr(\mathbf{f}_p | X_p = 0, \boldsymbol{\theta})}{\Pr(\mathbf{f}_p | X_p = 1, \boldsymbol{\theta})} = \sum_{p \in \Omega} x_p v_p = \mathbf{x}^t \mathbf{v} = \sum_{p \in S^1} v_p, \qquad (1.12)$$

where unary potential v_p is equal to the logarithm of likelihood ratio:

$$v_p = \ln \frac{\Pr(\mathbf{f}_p | X_p = 0, \boldsymbol{\theta})}{\Pr(\mathbf{f}_p | X_p = 1, \boldsymbol{\theta})},$$

and $\mathbf{v} = (v_p)_{p \in \Omega}$. Symbol $\stackrel{c}{=}$ denotes equality, up to an additive constant independent of binary segmentation variable $\mathbf{x} \in \{0, 1\}^{|\Omega|}$. Fig. 1.1 (b) depicts the image of unary potentials v_p: v_p is positive when the background likelihood at pixel p is higher than the foreground likelihood, i.e., $\Pr(\mathbf{f}_p | X_p = 0, \boldsymbol{\theta}) \geq \Pr(\mathbf{f}_p | X_p = 1, \boldsymbol{\theta})$, and negative otherwise. Now, minimizing the sum of unary potentials $\sum_{p \in \Omega} x_p v_p$ with respect of binary variables $x_p \in \{0, 1\}$ has a very clear meaning and is trivial. The global minimum can be obtained trivially by just setting $x_p = 1$ when v_p is negative, as this would decrease the sum, and $x_p = 0$ otherwise. This has a clear meaning: A pixel p is assigned to the foreground ($x_p = 1$), when $\Pr(\mathbf{f}_p | X_p = 1, \boldsymbol{\theta}) \geq \Pr(\mathbf{f}_p | X_p = 0, \boldsymbol{\theta})$, and to the background otherwise. In the literature, binary functions taking the form of a sum of unary potentials, as in Eq. (1.12), are also referred to as *linear* terms [11] because they are linear functions of segmentation variable $\mathbf{x}$. Fig. 1.1 (c) depicts the foreground region obtained by minimizing this sum of unary potentials. Even though we used "perfect" likelihood models (normalized color histograms) estimated from the ground truth, the obtained foreground region is noisy with several small, isolated regions. This example shows how unary potentials are not enough, even in this basic binary segmentation scenario with "perfect" knowledge of the regional probability models. Basic pairwise potentials, which we will discuss in the next section, can correct the result, yielding smooth (regularized) segmentation boundaries, as illustrated in Fig. 1.1 (d).

(a) Image   (b) Likelihood ratios v_p (c) Unary potentials   (d) Pairwise potentials

Figure 1.1 Binary segmentation. A basic segmentation example based on fitting given (fixed) regional likelihood models, which takes the form of unary potentials, and on a MRF regularization, which takes the form of pairwise potentials.

1.1.3 MRF pairwise potentials

In the general MRF function in Eq. (1.9), the second term can take the form of a sum of pairwise potentials, which have a regularization effect. Such a regularization embeds a prior on the fact that pixels (p, q), which are neighbors in a regular image grid, are likely to belong to the same segmentation region. A pairwise potential $\psi_{p,q}(x_p, x_q)$ evaluates the penalty of assigning a pair of labels $\{x_p, x_q\}$ to a pair of pixels $\{p, q\}$. It can be written in the following general form:

$$\psi_{p,q}(x_p, x_q) = w_{p,q}c(x_p, x_q), \tag{1.13}$$

where $c(x_p, x_q)$ is a label-compatibility function. The standard *Potts* model [12] is an important example of pairwise regularization of the form (1.13) and is a mainstay in computer vision. It uses the following basic label compatibility:

$$c(x_p, x_q) = [x_p \neq x_q], \tag{1.14}$$

where [.] denotes the Iverson bracket, taking value 1 if its argument is true and 0 otherwise. Let us consider the simple case of a binary segmentation, with pairwise potentials $w_{p,q}$ equal to a positive constant λ if p and q are neighbors (e.g., a 4-, 8-, or 16-neighborhood system), and to 0 otherwise. Minimization of such pairwise potentials encourages nearby pixels to have the same label: We pay penalty λ when two neighboring pixels do not have the same label (i.e., one belongs to the foreground and the other to the background) because, in this case, $[x_p \neq x_q] = 1$. In this binary-segmentation case, it is easy to see that the Potts term is proportional to the *length* of the segmentation boundary. Therefore, minimizing this term acts as a prior, which discourages small, isolated regions in the solution, and favors smooth (regular) segmentation boundaries; see the solution in Fig. 1.1 (d). This pairwise term is a particular choice of the MRF prior defined earlier in Eq. (1.5), with function $\mathcal{R}$ evaluating boundary length:

$$\Pr(\mathbf{X} = \mathbf{x}) \propto \exp\left(-\mathcal{R}(\mathbf{x})\right), \quad \text{with } \mathcal{R}(\mathbf{x}) = \lambda \sum_{p,q \in \mathcal{N}} [x_p \neq x_q], \tag{1.15}$$

where $\mathcal{N}$ denotes the set of pairs of neighboring pixels.

1.1.4 CRF pairwise potentials

A natural extension of the length prior in Eq. (1.15) is to introduce some image–data dependency. Instead of setting $w_{p,q}$ equal to a constant λ in the general pairwise form in (1.13), we can use $w_{p,q} = \lambda c_{p,q}(\mathbf{f})$:

$$\psi_{p,q}(x_p, x_q) = \lambda c_{p,q}(\mathbf{f})[x_p \neq x_q], \tag{1.16}$$

where the additional data-dependent function $c_{p,q}$ lessens the penalty that we pay (i.e., $\lambda c_{p,q}$) whenever image information implies the presence of a boundary, e.g., locations for which image contrast is high (image edges). One can choose, for instance, a decreasing function of image–feature difference $\|\mathbf{f}_p - \mathbf{f}_q\|$, which attracts the segment boundary towards strong image edges [6,8]:

$$c_{pq} \propto \exp\left(-\beta \|\mathbf{f}_p - \mathbf{f}_q\|^2\right), \tag{1.17}$$

where $\|.\|$ denotes the Euclidean distance. With the dependence of c_{pq} on image data $\mathbf{f}$, pairwise terms of the form (1.16) are often referred to as conditional random fields (CRFs). In general, similarly to MRFs, a CRF also uses a posterior modeling the latent variables of $\mathbf{X}$ given observations $\mathbf{f}$, i.e., $\Pr(\mathbf{X} = \mathbf{x}|\mathbf{f})$. However, the difference with MRF is that factorization of this posterior into a prior $\Pr(\mathbf{X} = \mathbf{x})$ and a likelihood $\Pr(\mathbf{f}|\mathbf{X} = \mathbf{x})$ is not done explicitly, unlike Eq. (1.6). This makes it convenient to embed some form of complex dependencies between the latent segmentation and data directly in the posterior, as done in the pairwise potentials in Eq. (1.16).

User inputs   Segmentation

Figure 1.2 Segmentation with user scribbles and edge-sensitive CRF regularization.

Fig. 1.2 shows an interactive segmentation example based on edge-sensitive pairwise potentials, which were used in conjunction with unary potentials defined from user-provided scribbles. The unary potentials include log-likelihoods based on color models estimated from the user-provided scribbles. In addition to the log-likelihoods, these unary potentials also contain large constant values for scribble pixels; these act as simple hard constraints to force scribble pixels to belong to their corresponding foreground or background regions.

Although very common in practice, it is worth noting that the Potts model does not account for semantic information. For instance, the penalty we pay for a pair of nearby variables labeled "car" and "road" is the same as the penalty for "car" and "cat". It is possible to use semantic compatibilities that enforce priors on the sets of labels (classes) appearing together in images, e.g., co-occurrence potentials [2,13]. One can also learn the compatibilities from training images with ground-truth segmentations [3].

1.1.5 Dense CRF

Fully connected pairwise potentials [3], often referred to as *dense CRF*, have attracted significant attention recently. For instance, in the context of deep semantic image segmentation [14–17], state-of-the-art results were obtained recently by integrating the popular dense CRF potentials of Krähenbühl and Koltun [3] and the unary scores of a convolutional neural network (CNN) classifier. The dense pairwise potentials in [3] take the form in Eq. (1.13), with $w_{p,q}$ expressed for each pair of pixels $(p, q) \in \Omega^2$ as the sum of two kernels, each of a Gaussian form:

$$w_{p,q} = w_2 \exp\left(-\frac{\|p - q\|^2}{2\sigma_1^2} - \frac{\|\mathbf{f}_p - \mathbf{f}_q\|^2}{2\sigma_2^2}\right) + w_1 \exp\left(-\frac{\|p - q\|^2}{2\sigma_3^2}\right).$$

Notice that, here, we have long-range interactions between pairs of pixels, unlike standard sparse (grid) random fields such as length regularization: Pairwise potentials $w_{p,q}$ have nonzero values for pixels that are not neighbors in spatial domain Ω. The first kernel is based on appearance. It encourages neighboring pixels with similar colors to have the same label. Parameters σ_1 and σ_2 control the amounts of neighborhood (how close the pixels are) and appearance similarity, respectively. The second kernel enforces smoothness. It discourages the occurrences of small and isolated regions in the solutions. The label compatibility function can be either Potts, as in Eq. (1.14), or other functions that account for semantic information, for example, using co-occurrence priors [2,13] or learning the compatibilities from training images [3].

A typical choice for unary potentials is the negative log-likelihoods, as in Eq. (1.10), where $\Pr(\mathbf{f}_p | X_p = x_p, \boldsymbol{\theta})$ corresponds to a generative probability model of image features (e.g., colors, textures, motion), within the image subdomain (region) having label l. This was the case of the popular Boykov–Jolly model [6,7], which we discussed earlier. Another option is to use the negative log-posteriors as unary penalties:

$$\psi_p(x_p) = -\ln \Pr(X_p = x_p | \mathbf{f}, \boldsymbol{\theta}). \qquad (1.18)$$

This is the case of state-of-the-art semantic segmentation techniques [14–16] based on deep learning for building posteriors (1.18). These methods embed the unary scores of a discriminative convolutional neural network (CNN) classifier into a CRF model. In this case, $\boldsymbol{\theta}$ are the parameters of the CNN, which are learned *a priori* from a large number of training images with ground-truth segmentations. The CRF regularization is used as a postprocessing step, which enhances substantially the performance of the CNN classifier, as in the case of the popular DeepLab model [15,16]. These works, among many other, showed that integrating the unary scores of a CNN classifier and the pairwise potentials of dense CRF achieve highly competitive performance. Popularized by the DeepLab [15,16], dense CRF has becomes the de facto choice for semantic CNN segmentation, both in the fully [14,16] and weakly/semi [17–20] supervised learning settings.

Many recent works showed experimentally that, when used in conjunction with modern, fully supervised and deep CNN classifiers, dense CRFs yield, typically, better performances than classical grid (or sparse) CRFs. Traditionally, grid CRFs, which connect only neighboring pixels, were deployed to regularize (or smooth) the boundaries of the segmentation regions (since they evaluate boundary length), and to align the predictions

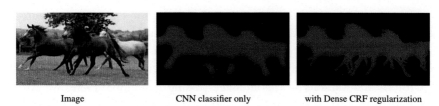

Image   CNN classifier only   with Dense CRF regularization

Figure 1.3 Semantic segmentation integrating deep CNNs and dense CRF regularization. The unary scores are the posteriors of a discriminative deep network, whose parameters are learned *a priori* from a large number of training images with ground-truth segmentations. The examples are from [15].

with image edges (in the case of edge-sensitive regularization). This is extremely useful when the unary potentials, either likelihoods or posteriors, are noisy, as is typically the case when using weak classifiers, e.g., those based on hand-crafted features such as colors. In general, such hand-crafted classifiers yield irregular boundaries and spurious predictions in the form of small, isolated regions; see the example in Fig. 1.1. On the contrary, when fully supervised with large amounts of training data,[3] deep CNN classifiers yield unary posteriors that correspond to smooth segmentation boundaries; see the example in Fig. 1.3. In this setting, it might be more appropriate to use dense rather than grid CRFs because the purpose is to recover local structures rather than boundary smoothing. Using grid CRF might lead to over-smoothing that misses thin structures.

1.2 Popular optimizers for random fields

Even when we limit ψ_c in the general-form CRF (1.9) to pairwise potentials, the ensuing discrete optimization problems are NP-hard [21]. While the literature on optimizing general CRF model (1.9) is very large, there are two types of optimization techniques that are dominant in computer vision: discrete combinatorial graph cuts [22] and variational mean-field inference [3]. These techniques have made a substantial impact in vision for tackling a wide class of pairwise conditional CRFs. We will dedicate the next chapters to discuss these optimization methods in greater detail. For graph cuts, we will discuss the popular Boykov–Kolmogorov [22] algorithm for binary segmentation, as well as the well-known expansion,

---

[3] Full supervision means the use of large amounts of images, each with the associated ground-truth segmentation built manually with human effort.

and swap moves for multi-region (or multi-label) problems [12]. In mean-field inference, we will discuss the parallel message-passing algorithm of Krähenbühl and Koltun [3], which has becomes the de facto choice in recent semantic segmentation models [14]. We will also examine various extensions and variants of this message-passing technique. Here, we will just give brief discussions of the pros and cons of these very popular optimization methods in regard to optimality guarantees and flexibility as well as computational scalability for computer vision problems with millions of latent variables, as is the case of semantic image segmentation.

1.2.1 Graph cuts

Graph cut is a dominant optimization technique in computer vision that has had substantial impact on solving discrete problems involving combinations of unary and pairwise *submodular* regularization potentials [12,22]. A function $\psi_{p,q}(x_p, x_q)$ defined over a pair of discrete binary variables x_p and x_q is submodular if and only if

$$\psi_{p,q}(1,0) + \psi_{p,q}(0,1) \geq \psi_{p,q}(1,1) + \psi_{p,q}(0,0). \qquad (1.19)$$

For instance, the Potts regularization examined earlier for binary segmentation is a submodular function. More generally, submodular functions were instrumental in the development of various computer vision algorithms for a broad swath of problems [1]. Such submodular functions are amenable to powerful combinatorial optimization algorithms, which provide excellent guarantees as to solution quality, while being computationally efficient in many practical instances in image analysis. In fact, it is even possible to obtain exact global optima for some special cases. For instance, the global optimum of a function containing unary and submodular pairwise potentials, as is the case of the binary segmentation examples discussed earlier, can be computed exactly in low-order polynomial time (with respect to the image size). This can be done with a graph cut by solving an equivalent max-flow problem [22]. In addition to global-optimality in the binary case, move-making algorithms based on graph cuts provide bounds on the solution quality in the multi-region (or multi-label) case [12]. Those optimization aspects will be discussed in greater details in Chapter 2.

In practice, it is well known that the popular Boykov–Kolmogorov (BK) algorithm [22] yields a state-of-the-art empirical performance in the context of 2D grids with sparse neighborhood systems (4-, 8-, or 16-neighborhood), as is the case of typical vision problems such as 2D segmentation with length regularization. The algorithm uses heuristics that handle efficiently sparse 2D grids. However, the efficiency of the algorithm may decrease when moving from 2D to 3D (or higher-dimensional) grids or when using denser (larger neighborhood) grids. Furthermore, distributing the computations for the BK algorithm[4] is not a trivial problem and is the subject of active research [23–26].

1.2.2 Mean-field inference

Unlike graph cuts, fully connected (dense) graphs can be efficiently addressed with *mean-field* variational inference [2,3]. Although the general principle of mean-field inference is quite old [1], this class of first-order optimization algorithms has recently attracted a substantial interest in computer vision, yielding state-of-the art performances in various computer vision tasks [2,13–15,27–29], while reducing significantly the computational load. Computationally efficient implementations of mean-field inference are facilitated by fast Gaussian filtering techniques, which are common in the signal processing community [30]. Recall that optimizing a CRF function $\mathcal{E}(\mathbf{x}|\mathbf{f}, \boldsymbol{\theta})$ of the general form (1.9) with respect to $\mathbf{x}$ is equivalent to finding the MAP estimate of the corresponding Gibbs distribution:

$$\hat{\mathbf{x}} = \text{argmax}_{\mathbf{x}} \, \text{Pr}(\mathbf{X} = \mathbf{x}|\mathbf{f}, \boldsymbol{\theta}),$$

where $\text{Pr}(\mathbf{X} = \mathbf{x}|\mathbf{f}, \boldsymbol{\theta}) \propto \exp(-\mathcal{E}(\mathbf{x}))$. The main idea of mean-field algorithms is to approximate this Gibbs distribution by a simpler distribution $U(\mathbf{X} = \mathbf{x})$, which makes it easier to compute the MAP estimate [1–3]. To do so, we need to specify the form of approximation $U(\mathbf{X} = \mathbf{x})$, a measure of divergence (or similarity) between distributions $\text{Pr}(\mathbf{X} = \mathbf{x}|\mathbf{f}, \theta)$ and $U(\mathbf{X} = \mathbf{x})$ and an optimization technique, with the latter seeking the best U minimizing the divergence measure. Mean-field inference techniques are approximate first-order optimizers. Therefore, they do not provide the same strong optimality guarantee as graph cuts. However, despite this limitation, they have become very popular recently due to many aspects. First, they are flexible with respect to the choice of pairwise potentials, which are not required to be submodular, while being amenable to parallelization and providing convergence guarantee [2]. Therefore, they can address efficiently expressive models such as Dense CRFs. Those optimization aspects will be discussed in greater details in Chapter 3.

---

[4] Augmenting-path, max-flow algorithms are based on global operations and, therefore, do not accommodate parallel/distributed implementations.

References

[1] A. Blake, P. Kohli, C. Rother, Markov Random Fields for Vision and Image Processing, MIT Press, 2011.

[2] P. Baqué, T.M. Bagautdinov, F. Fleuret, P. Fua, Principled parallel mean-field inference for discrete random fields, in: IEEE Conference on Computer Vision and Pattern Recognition (CVPR), 2016, pp. 5848–5857.

[3] P. Krähenbühl, V. Koltun, Efficient inference in fully connected CRFs with Gaussian edge potentials, in: Advances in Neural Information Processing Systems (NIPS), 2011, pp. 109–117.

[4] R. Achanta, A. Shaji, K. Smith, A. Lucchi, P. Fua, S. Süsstrunk, SLIC superpixels compared to state-of-the-art superpixel methods, IEEE Transactions on Pattern Analysis and Machine Intelligence 34 (11) (2012) 2274–2282.

[5] P. Kohli, L. Ladicky, P.H.S. Torr, Robust higher order potentials for enforcing label consistency, International Journal of Computer Vision 82 (3) (2009) 302–324.

[6] Y. Boykov, M.P. Jolly, Interactive graph cuts for optimal boundary and region segmentation of objects in n-d images, in: IEEE International Conference on Computer Vision (ICCV), 2001, pp. 105–112.

[7] Y. Boykov, G. Funka Lea, Graph cuts and efficient n-d image segmentation, International Journal of Computer Vision 70 (2) (2006) 109–131.

[8] C. Rother, V. Kolmogorov, A. Blake, Grabcut: interactive foreground extraction using iterated graph cuts, ACM Transactions on Graphics 23 (3) (2004) 309–314.

[9] A. Khoreva, R. Benenson, J.H. Hosang, M. Hein, B. Schiele, Simple does it: Weakly supervised instance and semantic segmentation, in: IEEE Conference on Computer Vision and Pattern Recognition (CVPR), 2017, pp. 1665–1674.

[10] D. Lin, J. Dai, J. Jia, K. He, J. Sun, Scribblesup: Scribble-supervised convolutional networks for semantic segmentation, in: IEEE Conference on Computer Vision and Pattern Recognition (CVPR), 2016, pp. 3159–3167.

[11] I. Ben Ayed, L. Gorelick, Y. Boykov, Auxiliary cuts for general classes of higher-order functionals, in: IEEE International Conference on Computer Vision and Pattern Recognition (CVPR), 2013, pp. 1304–1311.

[12] Y. Boykov, O. Veksler, R. Zabih, Fast approximate energy minimization via graph cuts, IEEE Transactions on Pattern Analysis and Machine Intelligence 23 (11) (2001) 1222–1239.

[13] V. Vineet, J. Warrell, P.H.S. Torr, Filter-based mean-field inference for random fields with higher-order terms and product label-spaces, International Journal of Computer Vision 110 (3) (2014) 290–307.

[14] A. Arnab, S. Zheng, S. Jayasumana, B. Romera-Paredes, M. Larsson, A. Kirillov, et al., Conditional random fields meet deep neural networks for semantic segmentation: Combining probabilistic graphical models with deep learning for structured prediction, IEEE Signal Processing Magazine 35 (1) (2018) 37–52.

[15] L.C. Chen, G. Papandreou, I. Kokkinos, K. Murphy, A.L. Yuille, Semantic image segmentation with deep convolutional nets and fully connected CRFs, in: International Conference on Learning Representations (ICLR), 2015, pp. 1–14.

[16] L. Chen, G. Papandreou, I. Kokkinos, K. Murphy, A.L. Yuille, Deeplab: Semantic image segmentation with deep convolutional nets, atrous convolution, and fully connected CRFs, IEEE Transactions on Pattern Analysis and Machine Intelligence 40 (4) (2018) 834–848.

[17] M. Rajchl, M.C.H. Lee, O. Oktay, K. Kamnitsas, J. Passerat-Palmbach, W. Bai, et al., Deepcut: Object segmentation from bounding box annotations using convolutional neural networks, IEEE Transactions on Medical Imaging 36 (2) (2017) 674–683.

[18] M. Tang, F. Perazzi, A. Djelouah, I. Ben Ayed, C. Schroers, Y. Boykov, On regularized losses for weakly-supervised CNN segmentation, in: European Conference on Computer Vision (ECCV), Part XVI, 2018, pp. 524–540.

[19] A. Kolesnikov, C.H. Lampert, Seed, expand and constrain: Three principles for weakly-supervised image segmentation, in: European Conference on Computer Vision (ECCV), Part IV, 2016, pp. 695–711.

[20] G. Papandreou, L. Chen, K.P. Murphy, A.L. Yuille, Weakly- and semi-supervised learning of a deep convolutional network for semantic image segmentation, in: IEEE International Conference on Computer Vision (ICCV), 2015, pp. 1742–1750.

[21] M. Li, A. Shekhovtsov, D. Huber, Complexity of discrete energy minimization problems, in: European Conference on Computer Vision (ECCV), Part II, 2016, pp. 834–852.

[22] Y. Boykov, V. Kolmogorov, An experimental comparison of min-cut/max-flow algorithms for energy minimization in vision, IEEE Transactions on Pattern Analysis and Machine Intelligence 26 (9) (2004) 1124–1137.

[23] J. Dolz, I. Ben Ayed, C. Desrosiers, DOPE: Distributed optimization for pairwise energies, in: IEEE Conference on Computer Vision and Pattern Recognition (CVPR), 2017, pp. 4095–4104.

[24] J. Liu, J. Sun, Parallel graph-cuts by adaptive bottom-up merging, in: IEEE Conference on Computer Vision and Pattern Recognition (CVPR), 2010, pp. 2181–2188.

[25] P. Strandmark, F. Kahl, Parallel and distributed graph cuts by dual decomposition, in: IEEE Conference on Computer Vision and Pattern Recognition (CVPR), 2010, pp. 2085–2092.

[26] A. Bhusnurmath, C.J. Taylor, Graph cuts via l_1 norm minimization, IEEE Transactions on Pattern Analysis and Machine Intelligence 30 (10) (2008) 1866–1871.

[27] M. Saito, T. Okatani, K. Deguchi, Application of the mean field methods to MRF optimization in computer vision, in: IEEE Conference on Computer Vision and Pattern Recognition (CVPR), 2012, pp. 1680–1687.

[28] S. Zheng, S. Jayasumana, B. Romera-Paredes, V. Vineet, Z. Su, D. Du, et al., Conditional random fields as recurrent neural networks, in: IEEE International Conference on Computer Vision (ICCV), 2015, pp. 1529–1537.

[29] Z. Liu, X. Li, P. Luo, C.C. Loy, X. Tang, Deep learning Markov random field for semantic segmentation, IEEE Transactions on Pattern Analysis and Machine Intelligence 40 (8) (2018) 1814–1828.

[30] A. Adams, J. Baek, M.A. Davis, Fast high-dimensional filtering using the permutohedral lattice, Computer Graphics Forum 29 (2) (2010) 753–762.

CHAPTER 2

Graph cuts

2.1 Min-cut and max-flow problems

As discussed in the previous chapter, standard conditional random field (CRF) models minimize functions of the following general form:

$$\mathcal{E}(\mathbf{x}) = \sum_{p \in \Omega} \psi_p(x_p) + \frac{1}{2} \sum_{p,q \in \Omega^2} \psi_{p,q}(x_p, x_q), \qquad (2.1)$$

where $\mathbf{x} = (x_p)_{p \in \Omega} \in \{0, \ldots, L-1\}^{|\Omega|}$ denotes a label assignment, Ω is the spatial image domain and $\{0, \ldots, L-1\}$ a finite set of discrete labels. Each variable x_p assigns one label among these to pixel p. For instance, in the context of image segmentation, labels may represent the target regions. In this section, let us take binary (two-region) segmentation as an example for optimizing the objective function (2.1) with a graph cut. In this case, $L = 2$ and assignment variable x_p is in $\{0, 1\}$, with $x_p = 1$ indicating that pixel p belongs to the "foreground region" and $x_p = 0$ means assigning p to the "background".

$\psi_p(x_p)$ is a unary potential that measures the individual penalty that we pay if label $x_p \in \{0, 1\}$ is assigned to pixel p. In the previous chapter, we discussed several typical choices of unary potentials, e.g., log-likelihoods. For instance, $\psi_p(1)$ may be chosen to reflect how well the color of pixel p fits a given (known) probability model of the foreground region (e.g., Gaussian mixture models). Also, $\psi_p(0)$ might be chosen in a similar way: It evaluates a fitting between the color of p and a given model of the background.

$\psi_{p,q}(x_p, x_q)$ is a pairwise potential that evaluates the penalty of assigning a pair of labels $\{x_p, x_q\}$ to a pair of pixels $\{p, q\}$. Potts regularization, which we discussed in great detail in Chapter 1, is a widely used pairwise function, namely,

$$\psi_{p,q}(x_p, x_q) = w_{p,q}[x_p \neq x_q], \qquad (2.2)$$

where, recall, [.] denotes the Iverson bracket, taking value 1 if its argument is true and 0 otherwise. Pairwise potentials $w_{p,q}$ are strictly positive if p and q are neighbors (e.g., using a 4-, 8-, or 16-neighborhood system),

High-Order Models in Semantic Image Segmentation
https://doi.org/10.1016/B978-0-12-805320-1.00007-5

and equal to 0 otherwise. In this case, each $w_{p,q}$ is a penalty for assigning different labels to neighboring pixels p and q. Such a penalty can be either a constant, in which case the regularization term measures the length of segment boundary, or a decreasing function of the difference (in norm) between the image features (e.g., color) of the two pixels, which attracts the segment boundary towards strong feature edges [3]. Therefore, the Potts model acts as a prior on segmentation boundary, encouraging the latter to be smooth (regular) and/or aligned with the image edges.

Potts regularization belongs to an important family of binary pairwise functions, i.e., *submodular* functions. A function $\psi_{p,q}(x_p, x_q)$ defined over a pair of binary variables x_p and x_q is submodular if and only if:

$$\psi_{p,q}(1, 0) + \psi_{p,q}(0, 1) \geq \psi_{p,q}(1, 1) + \psi_{p,q}(0, 0). \qquad (2.3)$$

More generally, beyond binary-variable functions, submodular functions have been instrumental in the development of various computer vision algorithms, for a spectrum of problems [4]. Such submodular functions are amenable to powerful combinatorial optimization algorithms, which provide excellent guarantees as to solution quality, while being computationally efficient. Even though optimization problems of the form (2.1) are NP-hard [5], it is possible to obtain exact global optima for some special cases. For binary variables, the global optimum of a function containing unary and submodular pairwise potentials can be computed exactly in low-order polynomial time (with respect to the image size), via a graph cut, by solving an equivalent max-flow problem [1]. In addition to global optimality in the binary case, *move-making* algorithms based on graph cuts provide highly competitive solutions and optimality bounds in the multi-label case [2], for discrete submodular functions of the general from in (2.1). Therefore, they have made a substantial impact in computer vision.

In the remainder of this section, we discuss in more detail how to cast optimizing function (2.1) in the binary-variable case ($L = 2$) such as finding the minimum-cut/maximum-flow of an S-T graph with two terminal nodes (S and T). Then, in Section 2.2, we will discuss in great details move-making algorithms for the multi-label case ($L > 2$).

2.1.1 S-T cuts

Consider optimizing objective (2.1) in the case of a binary variable $\mathbf{x} \in \{0, 1\}^{|\Omega|}$ and a pairwise function $\psi_{p,q}$ given by submodular Potts model (2.2).

To this end, one could construct a *weighted* graph [3,4]:

$$G = \langle \mathcal{V}, \mathcal{W} \rangle, \tag{2.4}$$

where $\mathcal{V}$ is a set of nodes (vertices) of the graph and $\mathcal{W}$ a set of undirected weighted edges connecting these nodes. $\mathcal{V}$ has a node for each pixel $p \in \Omega$, in addition to two virtual nodes, S and T, called *terminals*: $\mathcal{V} = \Omega \cup \{S, T\}$. Fig. 2.1 depicts an illustration of an S-T graph for optimizing (2.1) with respect to a binary variable, as in the case of two-region segmentation. In graph-cut/max-flow terminology, S is often referred to as the "source" terminal, and T as the "sink". In binary segmentation, S represents the foreground region, and T corresponds to the background (complement of the foreground). Beyond binary variables and two-region segmentation, the structure of the graph may vary. However, typically, in computer vision applications, graph nodes correspond to pixels, while terminals represent the labels. There are two types of edges in $\mathcal{W}$: p-links, which connect pixels in $\mathcal{N}$, with $\mathcal{N}$ denoting the set of pairs of neighbors in Ω (e.g., a 4-, 8-, or 16-neighborhood system); and t-links, which connect each pixel p to the source and sink terminals (S and T). Each edge is associated to some nonnegative weight. Typically, for optimizing functions of the form (2.1), the weights of the t-links are chosen so as to reflect the individual penalties that we pay for assigning a pixel to a label (i.e., unary potentials), whereas the weights of the p-links encode the pairwise regularization potentials.

An S-T cut $C \subset \mathcal{W}$ on a graph with two terminals is a subset of edges whose removal divides the graph into two disconnected subgraphs, each containing a terminal node. Fig. 2.1 illustrates how the graph ensuing from a cut, i.e., $\langle \mathcal{V}, \mathcal{W} \setminus C \rangle$, separates the terminals and partitions of the nodes into two disjoint subsets. A cut corresponds to a binary-variable labeling, as in the case of two-region segmentation. It assigns each node (pixel) p in Ω to one of the two terminals. In combinatorial optimization, the cost of a cut is defined as the sum of its edge weights. Furthermore, the globally minimum cut, i.e., the one that has the lowest cost among all possible cuts, could be computed in low-order polynomial time by solving an equivalent max-flow problem [1]. One could choose properly the weights of the edges of the graph, so that the minimum-cost cut induces a labeling that globally minimize the objective function. For instance, the edge weights in Table 2.1 correspond to optimizing (2.1) with pairwise function $\psi_{p,q}$ given by the Potts model (2.2).

The edge weights in Table 2.1 have a clear interpretation. Recall that $\psi_p(0)$ is the penalty we pay for assigning pixel p to the background (or ter-

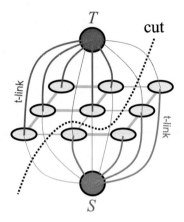

Figure 2.1 Illustration of an *S-T* cut on a graph for a binary segmentation of a 3 × 3 image, with a "source" terminal (*S*) representing the foreground region and a "sink" terminal (*T*) representing the background. Edge thickness, for both the *p*-links and *t*-links, reflects the weight (cost) of an edge. Low-weight edges are attractive choices for the minimum-cost cut, which induces a segmentation minimizing the objective function. Figure inspired by [3].

Table 2.1 Setting the weights of the *S-T* graph edges so that the cost of cut correspond to the objective function in (2.1) for the binary case, with pairwise function $\psi_{p,q}$ given by the Potts model (2.2).

$\{p, S\}$	$\{p, T\}$	$\{p, q\}$
$\psi_p(0)$	$\psi_p(1)$	$w_{p,q}$

minal node *T*), while $\psi_p(1)$ is the penalty for assigning *p* to the foreground (or terminal node *S*). Assume, without loss of generality, that $\psi_p(1) < \psi_p(0)$. In this case, the minimum–cost cut removes edge $\{p, T\}$ because it has a lower weight than edge $\{p, S\}$, thereby including *p* in the foreground region (as *p* remains connected to *S*). This makes sense because assigning label 1 to *p* favors the less expensive penalty, $\psi_p(1)$ and since it corresponds to a lower value of the minimized objective function. When $\psi_p(0) < \psi_p(1)$, the minimum–cost cut removes edge $\{p, S\}$, assigning the pixel to the background. The weight of the *p*-link edge connecting neighboring pixels *p* and *q* is $w_{p,q}$, which, in the case of the Potts model, is the pairwise penalty that we pay for assigning different labels to *p* and *q* (i.e., for disconnecting these two pixels). Therefore, low-weight edges are attractive choices for the minimum-cost cut.

2.1.2 Standard combinatorial algorithms

In combinatorial optimization, the minimum S-T cut could be found by solving an equivalent *max-flow* problem, i.e., finding the maximum flow from "source" terminal S to "sink" terminal T. Informally, a flow could be viewed as the "amount of water" that one sends from S to T, and edges could be viewed as "pipes", each having a capacity encoded by its weight. The theorem of Ford and Fulkerson [6] states that the maximum flow from terminal S to terminal T saturates a subset of graph edges, thereby partitioning the vertices of the graph into two subsets. Such a partition corresponds to the minimum cut, hence the equivalence between max-flow and min-cut problems. In fact, the minimum-cut cost is equal to the value of the maximum flow. There are standard combinatorial algorithms that enable providing min-cut and max-flow solutions on S-T graphs in polynomial time, including "augmenting-path" [6,7] and "push-relabel" [8] methods.

Augmenting-path methods push source-to-sink flows within the unsaturated paths of an S-T graph, up to a maximum value of the flow (saturation). The typical procedure for such methods is as follows. They track the distribution of the current flow $\mathcal{F}$ within the edges of a graph $\mathcal{G}$ via a residual graph $\mathcal{G}_r$. The latter has exactly the same topology as $\mathcal{G}$. However, the capacities of a given edge in $\mathcal{G}_r$ and $\mathcal{G}$ are different. In $\mathcal{G}_r$, for each edge, we store the residual capacity, given the current amount of flow that already exists within the edge. At initialization, $\mathcal{F} = 0$, i.e., there is no S-T flow, with the initial residual capacities of $\mathcal{G}_r$ set equal to those of original graph $\mathcal{G}$. At each iteration, an augmenting-path method seeks the shortest S to T path within unsaturated edges in $\mathcal{G}_r$. If a shortest path is found, its flow is increased by a maximum possible value $\Delta\mathcal{F}$, which saturates at least one edge within the path. Furthermore, all the residual capacities within the path are decreased by a value of $\Delta\mathcal{F}$. This yields an increase of the total S-to-T flow, from its current value $\mathcal{F}$ to $\mathcal{F} + \Delta\mathcal{F}$. An augmenting-path algorithm achieves the maximum flow when any S-to-T path has at least one saturated edge within $\mathcal{G}_r$.

An example of augmenting-path approaches is Dinic's algorithm [7], which seeks shortest S-to-T paths within graph $\mathcal{G}_r$, via a breadth-first search. Once all shortest S-to-T paths of length l reach saturation, the method starts from the scratch seeking S-to-T paths of length $l + 1$. The worst-case complexity of the method is $O(|\mathcal{W}||\mathcal{V}|^2)$, where $|\mathcal{W}|$ is the number of edges and $|\mathcal{V}|$ is the number of vertices. It is worth noting that, for such augmenting-path approaches, the use of the shortest paths is important for improving the theoretical complexity.

Push-relabel approaches [8] tackle max-flow problems in a different way because they do not track the distribution of the current flow. Instead, they use "active" nodes, which correspond to "flow excess", along with a labeling of nodes that correspond to a lower bound on the distance to the sink terminal. These methods proceed by "pushing" excess flows towards nodes that are within a small distance of the sink. Such "pushing" procedure is deployed for active nodes corresponding to the largest distance (label), and it yields progressive distance increases along with edge saturation. Flows that could not be delivered are returned back to the source terminal. For detailed discussions on both push-relabel and augmenting-path methods, we refer readers to [9].

2.1.3 The BK algorithm

Boykov–Kolmogorov (BK) algorithm [1] belongs to the family of augmenting-path methods and is widely used in computer vision. It has made a substantial impact on solving discrete problems involving pairwise submodular functions, which are common in spectrum of computer-vision applications. Based on heuristics that deal effectively with sparse grids (e.g., 2D grids with a 4-neighborhood system), the algorithm improves the empirical performance of the standard augmenting-path algorithms in computer vision applications.

Typically, when all paths of a given length have been explored, standard augmenting-path methods start a novel breadth-first search for S-to-T paths. However, in computer vision applications, using breadth-first search trees often requires exploring a large portion of the pixels in an image that could be prohibitively time consuming if done frequently. Moreover, in realistic computer-vision experiments, rebuilding search trees, as is standard in augmenting-path methods, performs poorly [1]. The BK algorithm uses two search trees for detecting augmenting paths, one from S to T and the other from T to S. However, an important difference with standard augmenting-path methods is that it does not create these trees from scratch at every step. One limitation of the BK algorithm is that the obtained augmenting paths do not necessarily correspond to the shortest ones, which invalidates the shortest-path complexity. For the BK algorithm, the augmentation number is upper bounded by the minimum-cut cost. Thus, the worst-case complexity of the BK algorithm is $O(|\mathcal{W}||\mathcal{V}|^2|C|)$ [1], where $\mathcal{V}$ is the number of nodes, $\mathcal{W}$ the number of edges and $|C|$ the cost of the optimal cut. This complexity is, in theory, worse than the standard max-flow algorithms. However, in practice, it is well known that the popular BK algorithm yields a state-of-the-art empirical performance in the context of sparse 2D grids [1], as is the case of typical computer-vision problems, outperforming significantly the standard max-flow algorithms previously mentioned. It is worth noting, though, that the efficiency of the algorithm may decrease when moving from 2D to 3D (or higher-dimensional) grids or when using denser (larger neighborhood) grids. Furthermore, similarly to other augmenting-path methods, the BK algorithm is based on global operations and, therefore, does not accommodate parallel/distributed implementations. Distributing the computations for the BK algorithm is not a trivial problem, and so it is the subject of active research [10–13].

2.2 Move-making algorithms for multi-label problems

The previous section discussed optimization of submodular functions of the form (2.1) in the case of binary variables $\mathbf{x} \in \{0, 1\}^{|\Omega|}$, i.e., there are only two labels ($L = 2$). Of course, a wide range of segmentation tasks and, more generally, of computer vision problems needs multi-label variables ($L > 2$). In general, multi-label problems of the form (2.1) are difficult, and one requires approximate solutions for tackling them. Move-making algorithms are very popular in computer vision [2,4]. Under certain conditions, they provide powerful approximate solvers, along with optimality bounds, for multi-label problems of the form (2.1). Such move-making approaches are based on splitting the original multi-label problem into a sequence of binary-variable subproblems, each solved with a graph cut. Within each iteration, a move-making approach pursues two possible choices for a node p in a graph: Either we keep the current label, or we change it to a new label. Interestingly, this could be stated as a binary-variable optimization. For instance, one could associate each node p to a binary variable $t_p \in \{0, 1\}$, where $t_p = 0$ indicates that the current label of p is kept and $t_p = 1$ indicates that node p is switching to some other label. Hence, the purpose is the compute the "best" subset of nodes for which we need to change labels. What is meant by "best" here is the largest possible decrease of the overall objective. Of course, there are diverse choices as to the type of moves that are allowed during each iteration, and the difficulty of the ensuing binary-variable problems (i.e., optimization with respect to move-making variables t_p) depends on such choices. Interestingly, there are certain types of moves for which the ensuing binary problems are submodular and, therefore, could be solved effectively with a graph cut, as discussed in the previous section. In the following, we will discuss in more details *expansion* and *swap* moves [2], which are widely used in computer vision [4].

The general idea of move-making approaches is as follows: Given current labeling $\mathbf{x}$, we can move to another labeling $\mathbf{x}'$ only if the change is within a set of allowed moves. Such a set should be, ideally, defined in a way that enables to compute efficiently the best move (i.e., a move that yields the best decrease in the objective). Given an initial labeling, move-making approaches explore the best possible moves at each iteration and stop when there is no move that yields a decrease in the objective function.

2.2.1 Expansion and swap moves

Given a label α in $\{0, \ldots, L-1\}$, an α-expansion move from a current labeling $\mathbf{x} = (x_p)_{p \in \Omega}$ to a new one $\mathbf{x}' = (x'_p)_{p \in \Omega}$ satisfies the following condition:

$$\text{If } x_p \neq x'_p, \text{ then } x'_p = \alpha.$$

This means that allowed moves are only those that enable subsets of nodes to switch from their current labels to α.

Given any pair of labels α and β, an α-β swap is a move from $\mathbf{x}$ to $\mathbf{x}'$ that satisfies the following condition:

$$\text{If } x_p \neq x'_p, \text{ then } (x_p, x'_p) \in \{\alpha, \beta\}^2.$$

This means that allowed moves are only those that enable subset of nodes to switch their labels from α to β, or vice versa.

2.2.2 Structure of move-making algorithms

Move-making algorithms are based on the following iterative steps.

The case of α-β swaps:
1. Initialize the labeling to some arbitrary value $\mathbf{x}$
2. $s \leftarrow 0$
3. For each label pair $(\alpha, \beta) \in \{0, \ldots, L-1\}^2$,
   a. Solve the following problem:

$$\tilde{\mathbf{x}} = \arg\min_{\mathbf{x}'} \mathcal{E}(\mathbf{x}') \quad \text{s.t.} \quad \mathbf{x}' \text{is within one } \alpha\text{-}\beta \text{ swap of } \mathbf{x}$$

    b. If $\mathcal{E}(\tilde{\mathbf{x}}) < \mathcal{E}(\mathbf{x})$, then $\mathbf{x} \leftarrow \tilde{\mathbf{x}}$ and $s \leftarrow 1$
4. If $s = 1$, go to step 2.
5. Return $\mathbf{x}$

The case of α-expansion moves:
1. Initialize the labeling to some arbitrary value $\mathbf{x}$
2. $s \leftarrow 0$
3. For each label $\alpha \in \{0, \ldots, L-1\}$,
    a. Solve the following problem:

$$\tilde{\mathbf{x}} = \arg\min_{\mathbf{x}'} \mathcal{E}(\mathbf{x}') \quad \text{s.t.} \quad \mathbf{x}' \text{ is within one } \alpha\text{-expansion of } \mathbf{x}$$

    b. If $\mathcal{E}(\tilde{\mathbf{x}}) < \mathcal{E}(\mathbf{x})$, then $\mathbf{x} \leftarrow \tilde{\mathbf{x}}$ and $s \leftarrow 1$
4. If $s = 1$, go to step 2.
5. Return $\mathbf{x}$

Clearly the two variants of these move-making algorithms follow the same general structure, except that the allowed moves are different. An iteration corresponds to one execution of steps 3.a and 3.b, and a cycle corresponds to the three steps from 2 to 4. Within each cycle, we perform one iteration for every label pair in the case of swap moves and for every label in the case of expansion moves, following a certain order. The latter could be either random or fixed. A cycle is considered successful ($s \leftarrow 1$) when a solution that strictly decreases the objective function is obtained at any iteration. The algorithm attains termination when we reach the first unsuccessful cycle, i.e., one that does not decrease strictly the objective. Clearly, an expansion-move cycle takes L iterations, whereas a swap-move one requires L^2 iterations. These move-making algorithms are guaranteed to stop after a finite number of cycles. In fact, when functions ψ_p and $\psi_{p,q}$ are independent of the cardinality of the set of vertices of the graph (i.e., $|\Omega|$), one could demonstrate the stopping occurs within $O(|\Omega|)$ [4]. Experimentally, though, it is often observed that termination occurs within a few cycles, with the most significant decrease of the objective typically happening during the first cycle.

2.2.3 Finding the best moves

Clearly, the most important part is the step in 3.a that aims to find the best move within a set of allowed moves. The seminal work in [2] showed how to convert expansion and swap moves into binary-variable problems that are submodular. Therefore, one could use S-T graphs and the effective graph–cut/max–flow algorithms discussed in the previous section for the binary case.

In the following, we detail how expansion and swap moves could be represented with a binary variable $\mathbf{t} = (t_p)_{p \in \Omega} \in \{0, 1\}^{|\Omega|}$. For a given move-making algorithm, let $T(\mathbf{x}, \mathbf{t})$ denotes a label-transformation function, which takes current labeling $\mathbf{x}$ and binary-variable move $\mathbf{t}$ as inputs and outputs the labeling ensuing from the move, which we denote $\mathbf{x}'$: $\mathbf{x}' = T(\mathbf{x}, \mathbf{t})$. For α-expansion moves, the transformation should encode the fact that a given node either switches to label α or keeps its current label. Therefore, it could be written as follows:

$$x'_p = T_\alpha(x_p, t_p) = \begin{cases} \alpha & \text{if } t_p = 0, \\ x_p & \text{if } t_p = 1. \end{cases} \tag{2.5}$$

For α-β swaps, the transformation allows a node to change its current label α to β, or vice versa. Therefore, it could be encoded with a binary variable as follows:

$$x'_p = T_{\alpha\beta}(x_p, t_p) = \begin{cases} x_p & \text{if } x_p \neq \alpha \text{ and } x_p \neq \beta, \\ \alpha & \text{if } x_p = \alpha \text{ or } \beta \text{ and } t_p = 0, \\ \beta & \text{if } x_p = \alpha \text{ or } \beta \text{ and } t_p = 1. \end{cases} \tag{2.6}$$

The binary-variable objective function corresponding to a move $\mathbf{t} = (t_p)_{p \in \Omega} \in \{0, 1\}^{|\Omega|}$ is: $\mathcal{E}(\mathbf{x}') = \mathcal{E}(T(\mathbf{x}, \mathbf{t})) = \mathcal{E}_T(\mathbf{t})$. Hence, steps 3.a in the cited move-making algorithms are performed by optimizing globally $\mathcal{E}_T$ with respect to binary variable $\mathbf{t}$, via a graph cut, yielding the best decrease of multi-label objective $\mathcal{E}$.

For CRF functions of the form in (2.1), the binary-variable objective corresponding swap moves $\mathbf{t}$ could be written as follows:

$$\mathcal{E}_{T_{\alpha\beta}}(\mathbf{t}) = \sum_{p \in \Omega} \psi_p(T_{\alpha\beta}(x_p, t_p)) + \frac{1}{2} \sum_{p, q \in \Omega^2} \psi_{p,q}(T_{\alpha\beta}(x_p, t_p), T_{\alpha\beta}(x_q, t_q)). \tag{2.7}$$

Here, $(x_p)_{p \in \Omega}$ is fixed and $(t_p)_{p \in \Omega}$ is the binary optimization variable. In order for us to use a graph cut to minimize globally (2.7) with respect to binary variable $(t_p)_{p \in \Omega}$, the pairwise function in (2.7) should be submodular, i.e., $\psi_{p,q}(T_{\alpha\beta}(x_p, 1), T_{\alpha\beta}(x_q, 0)) + \psi_{p,q}(T_{\alpha\beta}(x_p, 0), T_{\alpha\beta}(x_q, 1)) \geq \psi_{p,q}(T_{\alpha\beta}(x_q, 1), T_{\alpha\beta}(x_q, 1)) + \psi_{p,q}(T_{\alpha\beta}(x_q, 0), T_{\alpha\beta}(x_q, 0))$.

Therefore, given the transformation expression in (2.6), the binary-variable objective in (2.7) is submodular when pairwise functions $\psi_{p,q}$ verify the following condition:

$$\psi_{p,q}(\beta, \alpha) + \psi_{p,q}(\alpha, \beta) \geq \psi_{p,q}(\beta, \beta) + \psi_{p,q}(\alpha, \alpha) \, \forall \alpha, \beta. \tag{2.8}$$

For expansion moves, the binary-variable objective reads

$$\mathcal{E}_{T_\alpha}(\mathbf{t}) = \sum_{p \in \Omega} \psi_p(T_\alpha(x_p, t_p)) + \frac{1}{2} \sum_{p,q \in \Omega^2} \psi_{p,q}(T_\alpha(x_p, t_p), T_\alpha(x_q, t_q)). \tag{2.9}$$

In order for $\mathcal{E}_{T_\alpha}$ to be submodular and given transformation expression (2.5), it suffices that pairwise functions $\psi_{p,q}$ verify the following condition:

$$\psi_{p,q}(\beta, \alpha) + \psi_{p,q}(\alpha, \gamma) \geq \psi_{p,q}(\beta, \gamma) + \psi_{p,q}(\alpha, \alpha) \, \forall \alpha, \beta, \gamma. \tag{2.10}$$

Note that the satisfaction of condition (2.10) implies that condition (2.8) is also verified. This means that the condition for using expansion moves is more restrictive than the one required for swap moves. Conditions (2.8) and (2.10) are, in fact, related to well-known definitions in metric spaces. A pairwise function $\psi_{p,q}$ is a metric if it satisfies the following conditions $\forall \alpha, \beta, \gamma$:

$$\psi_{p,q}(\alpha, \beta) \geq 0 \tag{2.11}$$

$$\psi_{p,q}(\alpha, \beta) = 0 \Leftrightarrow \alpha = \beta \tag{2.12}$$

$$\psi_{p,q}(\alpha, \beta) = \psi_{p,q}(\beta, \alpha) \tag{2.13}$$

$$\psi_{p,q}(\alpha, \beta) \leq \psi_{p,q}(\alpha, \gamma) + \psi_{p,q}(\gamma, \beta). \tag{2.14}$$

Generalizations of a metric include semimetric, when the triangular inequality in (2.14) is omitted, and quasi-metric, when the symmetry in (2.13) is omitted. Note that the condition for expansion moves in (2.10) holds when pairwise function $\psi_{p,q}$ is a metric or quasi-metric, whereas the condition for swap moves in (2.8) holds for both metrics, quasi-metrics or semimetrics. Many pairwise functions used in computer vision problems are metrics or semimetrics [4]. For instance, the pairwise function of the popular Potts model, which is given by $\psi_{p,q}(\alpha, \beta) = w_{p,q}[\alpha \neq \beta]$, is a metric.

2.2.4 Local minima and optimality bounds

In discrete optimization, the notion of a local minimum is different from the standard definition of a local minimum in continuous optimization. Typically, a discrete labeling $\tilde{\mathbf{x}}$ is considered a local minimum of an objective

$\mathcal{E}$ when it satisfies the following condition:

$$\mathcal{E}(\tilde{\mathbf{x}}) \le \mathcal{E}(\mathbf{x}) \,\forall \mathbf{x} \text{ "close to" } \tilde{\mathbf{x}}, \qquad (2.15)$$

where "$\mathbf{x}$ is close to $\tilde{\mathbf{x}}$" means that discrete labeling $\mathbf{x}$ is within a *single* move from $\tilde{\mathbf{x}}$. Of course, the type of moves that are allowed for an algorithm affects the quality of the obtained local minima. There are many discrete optimization algorithms that use "standard" moves, in which only one node is allowed to change its label [4]. This means that, for a local minimum with respect to standard moves, the objective could not be improved by changing the label of a single node. This is, in fact, a weak condition imposed on local minima that typically yields poor solutions for objectives of the form in (2.1). The expansion and swap methods previously discussed generate minima of objective (2.1) with respect to much larger moves. Unlike standard moves, expansion and swap moves enable changing simultaneously the labels of much larger numbers of nodes. Such large moves yield more substantial changes in the current solution, which might help avoiding poor local minima, while improving convergence rates. However, the number of possible labelings within a single expansion or swap move from a given $\mathbf{x}$ is exponentially large, which makes finding local minima satisfying (2.15) much more difficult than in the case of standard moves. In fact, a naive search of a local minimum with respect to expansion and swap moves would require exponential time, whereas searching for a local minimum with respect to standard moves is linear. Earlier in this chapter, we discussed in detail how to explore effectively the best move within expansion and swap moves. This could be done with a graph cut, by converting expansion and swap moves into submodular, binary-variable problems, following the seminal work in [2]. The optimization difficulty ensuing from allowing large moves comes with a reward in terms of optimality guarantees (bounds). In fact, it is possible to show that the local minimum achieved by expansion moves is guaranteed to be within a known multiplicative factor from the global minimum [2], and such a factor is as small as 2 for the popular Potts model.

Given an objective $\mathcal{E}$ of the form in (2.1), let $\tilde{\mathbf{x}}$ denote a local minimum when expansion moves are allowed, and let $\mathbf{x}^*$ denote the global minimum. The authors of [2] showed that, when pairwise function $\psi_{p,q}$ is a metric, the local minimum is within a known factor of the global minimum:

$$\mathcal{E}(\tilde{\mathbf{x}}) \le 2\,c\,\mathcal{E}(\mathbf{x}^*), \qquad (2.16)$$

where c depends of the pairwise function and is given by:

$$c = \max_{p,q} \left(\frac{\max_{\alpha \neq \beta} \psi_{p,q}(\alpha, \beta)}{\min_{\alpha \neq \beta} \psi_{p,q}(\alpha, \beta)} \right). \tag{2.17}$$

Clearly, for the popular Potts model, i.e., $\psi_{p,q}(\alpha, \beta) = w_{p,q}[\alpha \neq \beta]$, $c = 1$, and we have an optimality bound $\mathcal{E}(\tilde{\mathbf{x}}) \leq 2 c \mathcal{E}(\mathbf{x}^*)$. It is worth noting that, in practice, the results obtained by expansion algorithms are, typically, closer to the global optima than this theoretical bound [4].

References

[1] Y. Boykov, V. Kolmogorov, An experimental comparison of min-cut/max-flow algorithms for energy minimization in vision, IEEE Transactions on Pattern Analysis and Machine Intelligence 26 (9) (2004) 1124–1137.

[2] Y. Boykov, O. Veksler, R. Zabih, Fast approximate energy minimization via graph cuts, IEEE Transactions on Pattern Analysis and Machine Intelligence 23 (11) (2001) 1222–1239.

[3] Y. Boykov, G. Funka Lea, Graph cuts and efficient n-d image segmentation, International Journal of Computer Vision 70 (2) (2006) 109–131.

[4] A. Blake, P. Kohli, C. Rother, Markov Random Fields for Vision and Image Processing, MIT Press, 2011.

[5] M. Li, A. Shekhovtsov, D. Huber, Complexity of discrete energy minimization problems, in: European Conference on Computer Vision (ECCV), Part II, 2016, pp. 834–852.

[6] L. Ford, D. Fulkerson, Flows in Networks, Princeton University Press, 1962.

[7] E.A. Dinic, Algorithm for solution of a problem of maximum flow in networks with power estimation, Soviet Mathematics. Doklady (1970).

[8] A.V. Goldberg, R.E. Tarjan, A new approach to the maximum-flow problem, Journal of the Association for Computing Machinery 35 (4) (1988) 921–940.

[9] W.J. Cook, W.H. Cunningham, W.R. Pulleyblank, A. Schrijver, Combinatorial Optimization, Wiley-Interscience, 1998.

[10] J. Dolz, I. Ben Ayed, C. Desrosiers, DOPE: Distributed optimization for pairwise energies, in: IEEE Conference on Computer Vision and Pattern Recognition (CVPR), 2017, pp. 4095–4104.

[11] J. Liu, J. Sun, Parallel graph-cuts by adaptive bottom-up merging, in: IEEE Conference on Computer Vision and Pattern Recognition (CVPR), 2010, pp. 2181–2188.

[12] P. Strandmark, F. Kahl, Parallel and distributed graph cuts by dual decomposition, in: IEEE Conference on Computer Vision and Pattern Recognition (CVPR), 2010, pp. 2085–2092.

[13] A. Bhusnurmath, C.J. Taylor, Graph cuts via l_1 norm minimization, IEEE Transactions on Pattern Analysis and Machine Intelligence 30 (10) (2008) 1866–1871.

CHAPTER 3

Mean-field inference

3.1 Pairwise conditional random field functions

Let us recall the pairwise conditional random field (CRF) functions we examined in previous chapters. Let $\mathbf{X} = (X_p)_{p \in \Omega}$ denotes $|\Omega|$ random variables, each associated with a pixel p, with Ω the spatial image domain. X_p takes its values in a finite set of discrete labels $l = \{0, \ldots, L-1\}$. In semantic segmentation of color images, for instance, each label corresponds a semantic category, e.g., "person", "street", "tree", "car", etc. Let $\mathbf{x} = (x_p)_{p \in \Omega}$ denotes a labeling of the image, i.e., a realization of random variable $\mathbf{X}$, which assigns a label $x_p \in \{0, \ldots, L-1\}$ to each pixel p. As discussed in Chapter 1, a CRF model seeks an optimal labeling that minimizes a pairwise function taking the following general form:

$$E(\mathbf{x}) = \sum_{p \in \Omega} \psi_p(x_p) + \frac{1}{2} \sum_{p,q \in \Omega^2} \psi_{p,q}(x_p, x_q). \tag{3.1}$$

Unary potential $\psi_p(x_p)$ measures the penalty that we pay if label x_p is assigned to pixel p. A typical choice is to use log-likelihoods:

$$\psi_p(x_p) = -\ln \Pr(\mathbf{f}_p | X_p = x_p, \boldsymbol{\theta}).$$

These correspond to generative probability models of pixel features $\mathbf{f}_p$ within the segmentation region, with $\boldsymbol{\theta}$ the parameters of the models. Such a choice is the case, for instance, of the popular Boykov–Jolly model [3,4]. Another option is to use log-posteriors:

$$\psi_p(x_p) = -\ln \Pr(X_p = x_p | \mathbf{f}, \boldsymbol{\theta}).$$

This is the case of state-of-the-art semantic segmentation techniques [2]. These integrate the unary scores of a discriminative convolutional neural network (CNN) classifier into a dense CRF of the form in Eq. (3.1), with $\boldsymbol{\theta}$ the network parameters, learned *a priori* from training data. Here, $\mathbf{f} = (f_p)_{p \in \Omega} \in \mathbb{R}^{|\Omega|}$ denotes the input image. This dense CRF postprocessing step enhances significantly the performance of CNN segmentation, as is the case of the popular DeepLab model [5,6].

High-Order Models in Semantic Image Segmentation
https://doi.org/10.1016/B978-0-12-805320-1.00008-7

Pairwise potential $\psi_{p,q}(x_p, x_q)$ evaluates the penalty of assigning pair of labels $\{x_p, x_q\}$ to pair of pixels $\{p, q\}$. It takes the general form of

$$\psi_{p,q}(x_p, x_q) = w_{p,q}c(x_p, x_q), \tag{3.2}$$

where $c(x_p, x_q)$ is a label-compatibility function. The Potts model, which we discussed in great details in Chapter 1, is the most basic form of label compatibility:

$$c(x_p, x_q) = [x_p \neq x_q].$$

Recall that [.] denotes the Iverson bracket, taking value 1 if its argument is true and 0 otherwise. For example, the length of segmentation boundary could be encoded with Potts, by setting potentials $w_{p,q}$ equal to a positive constant λ if p and q are neighbors, and to 0 otherwise. Minimizing a pairwise length term encourages regularized (smooth) segmentation boundaries.

In general, optimization problems of the form in Eq. (3.1) are NP-hard [7]. The previous chapter discussed in great detail graph-cut algorithms [8,9], which have made a substantial impact in computer vision because they can find exact global optima, or provide optimality bounds, for several special cases of pairwise potentials (3.2). In practice, it is well known that graph-cut algorithms yield state-of-the-art performances for typical vision problems with sparse and submodular pairwise potentials, e.g., length regularization. However, their efficiency may decrease when moving from 2D to 3D (or higher-dimensional) grids or when using denser (larger neighborhood) grids. There are many cases of practical interest of pairwise potentials (3.2) that cannot be handled effectively with graph cuts. For instance, dense CRF [1] regularization, which we discussed in great detail in Chapter 1, is based on fully connected pairwise potentials. In this case, we have long-range interactions between pairs of pixels, unlike standard sparse (grid) random fields such as length regularization: Pairwise potentials $w_{p,q}$ have nonzero values for pixels that are not neighbors in spatial domain Ω. Furthermore, the label compatibility function is not necessarily limited to the Potts model. It could be a more expressive function that accounts for semantic information, for example, using co-occurrence priors [10,11] or learning the compatibilities from training images [1]. Therefore, it involves a more general, hence more expressive, class of pairwise regularizers than standard length regularization. Recently, it has attracted significant attention, with excellent performances in the context of semantic image segmentation [2,6,12]. Unlike graph cuts, dense (fully connected) and/or nonsubmodular graphs can be efficiently addressed with mean-field variational inference [1,11].

The general principle of mean-field inference is quite old [13], and does not provide strong optimality guarantee. Nonetheless, recently, this class of first-order optimization algorithms has become very popular in computer vision [2,5,10,11,14–16], for several reasons. First, they are flexible with respect to the choice of pairwise potentials. Second, they are amenable to parallel, computationally efficient implementations, while providing convergence guarantee [11]. Therefore, they can address efficiently expressive models such as dense CRFs. Those optimization aspects will be discussed in detail in the following.

3.2 Mean-field inference

3.2.1 The mean-field approximation

Recall that optimizing function (3.1) is equivalent to finding the maximum *a posteriori* estimate (MAP) of the corresponding Gibbs distribution:

$$\hat{\mathbf{x}} = \arg\max_{\mathbf{x}} \Pr(\mathbf{X} = \mathbf{x}|\mathbf{f}, \theta), \tag{3.3}$$

with

$$\Pr(\mathbf{X} = \mathbf{x}|\mathbf{f}, \theta) = \frac{1}{Z}\exp(-E(\mathbf{x})) \tag{3.4}$$

and Z a normalization constant.

The main idea of mean-field algorithms is to approximate the Gibbs distribution in Eq. (3.4) by a simpler distribution $U(\mathbf{X} = \mathbf{x})t$, which makes it easier to compute the MAP estimate [1,11,13]. To do so, we need to specify the form of approximation $U(\mathbf{X} = \mathbf{x})$, a measure of divergence (or similarity) between distributions $\Pr(\mathbf{X} = \mathbf{x}|\mathbf{f}, \theta)$ and $U(\mathbf{X} = \mathbf{x})$ and an optimization technique, which seeks the best U minimizing the divergence measure. Typically, in the mean-field approximation, $U(\mathbf{X} = \mathbf{x})$ is assumed to be the product of the independent marginals:

$$U(\mathbf{X}) = \prod_{p \in \Omega} U_p(X_p = x_p). \tag{3.5}$$

Each marginal $U_p(X_p = x_p)$ is within the L-dimensional probability simplex:

$$\nabla_L = \{\mathbf{y} \in [0, 1]^L \mid \mathbf{1}^t\mathbf{y} = 1\}.$$

This means that it verifies the following constraints:

$$U_p(X_p = l) \geq 0 \quad \forall p, l$$
$$\sum_l U_p(X_p = l) = 1 \quad \forall p. \tag{3.6}$$

To keep the notation simple in the following, we use $u_{p,l}$ to denote $U_p(X_p = l)$, the probability of having label l at pixel p. Once U is obtained, the MAP estimate of $\Pr(\mathbf{X} = \mathbf{x}|\mathbf{f}, \theta)$ can be approximated by the MAP estimate of U. As U is fully factorized, the corresponding MAP estimate is just the label maximizing each independent marginal:

$$\hat{x}_p = \arg\max_l u_{p,l}.$$

Typically, the similarity measure is the Kullback–Leibler (KL) divergence:

$$\mathrm{KL}(U \| \Pr(.|\mathbf{f}, \theta)) = \sum_{p,l} u_{p,l} \log u_{p,l} + \sum_{p,l} v_{p,l} u_{p,l} + \frac{1}{2} \sum_{p,q} \sum_{l,l'} w_{p,q} c(l, l') u_{p,l} u_{q,l'}.$$
$$\tag{3.7}$$

Using convenient matrix and vector notations, minimizing the KL divergence in Eq. (3.7), subject to the simplex constraints, can be written as follows:

$$\min_{\mathbf{u}} \mathcal{E}(\mathbf{u}) \quad \text{s.t} \quad \mathbf{u}_p \in \nabla_L \, \forall p \in \Omega,$$

where:

$$\mathcal{E}(\mathbf{u}) = \mathbf{u}^t \ln \mathbf{u} + \mathbf{u}^t \mathbf{v} + \frac{1}{2} \mathbf{u}^t \Psi \mathbf{u}, \tag{3.8}$$

and we have the following notations:

- $\mathbf{u} \in [0, 1]^{L|\Omega|}$ takes the form $(\mathbf{u}_p)_{p \in \Omega}$, i.e., it concatenates simplex vectors $\mathbf{u}_p \in [0, 1]^L$, each containing the probability variables of all labels for pixel p: $\mathbf{u}_p = (u_{p,1}, \ldots, u_{p,L})^t$. $Ln\mathbf{u}$ has the same dimension as $\mathbf{u}$, with the logarithm applied to $\mathbf{u}$ component-wise.
- $\Psi = W \otimes C$ is the Kronecker product between the $|\Omega|$ by $|\Omega|$ matrix of pairwise potentials $W = [w_{p,q}]$ and an L by L label compatibility matrix $C = [c(l, l')]$. In the case of the standard Potts model, all the diagonal elements of C are equal to zero and off-diagonal elements to 1: $c_{l,l} = 0$ and $c_{l,l'} = 1$ for $l \neq l'$.
- $\mathbf{v} \in \mathbb{R}^{L|\Omega|}$ takes the form $(\mathbf{v}_p)_{p \in \Omega}$, with each $\mathbf{v}_p \in \mathbb{R}^L$ containing the unary potentials for pixel p: $\mathbf{v}_p = (v_{p,1}, \ldots, v_{p,L})^t$.

One of the main advantages of mean-field algorithms [11,17] is that they can accommodate parallel implementations while guaranteeing convergence to a local minimum of (3.8). This makes them practical for realistically-sized computer vision problems when there are substantial (dense) interactions between all pairs of variables. Such parallel and convergent implementations can be interpreted as bound optimization techniques, which guarantee that function (3.8) decreases when updating all the variables of **u** in parallel.

3.2.2 Bound optimization

Bound optimizers iteratively minimize an auxiliary function that is a tight upper bound of the original function. At iteration i, $\mathcal{A}_i(\mathbf{u})$ is an auxiliary function of $\mathcal{E}(\mathbf{u})$ at current solution $\mathbf{u}^i$ if it satisfies the following two conditions:

$$\mathcal{E}(\mathbf{u}) \le \mathcal{A}_i(\mathbf{u})$$
$$\mathcal{E}(\mathbf{u}^i) = \mathcal{A}_i(\mathbf{u}^i). \tag{3.9}$$

At each iteration, consider the following update of the current solution to the minimum of the auxiliary function:

$$\mathbf{u}^{i+1} = \arg\min_{\mathbf{u}} \mathcal{A}_i(\mathbf{u}).$$

These updates guarantee that, after each iteration, original function $\mathcal{E}$ does not increase:

$$\mathcal{E}(\mathbf{u}^{i+1}) \le \mathcal{A}_i(\mathbf{u}^{i+1}) \le \mathcal{A}_i(\mathbf{u}^i) = \mathcal{E}(\mathbf{u}^i).$$

The first inequality and last equality come from the auxiliary-function conditions in (3.9), i.e., $\mathcal{A}_i$ is a tight upper bound on the original function. The inequality in the middle comes from the fact that $\mathbf{u}^{i+1}$ is the minimum of $\mathcal{A}_i(u)$. Of course, optimizing the auxiliary function should be easier than the original problem. In the following, we will examine auxiliary functions of (3.8), whose optimization can be performed by parallel (independent) updates of each $u_{p,l} \, \forall p \in \Omega$. This is important for mean-field inference to scale up to realistic-size computer vision problems where $|\Omega|$ is of the order of millions of pixels for high-resolution images. First, we will examine auxiliary functions for the case Ψ is negative semi-definite, which means the pairwise term in Eq. (3.8) is concave. This is the case of the standard Potts model. Then, we will consider the general case of a symmetric matrix Ψ.

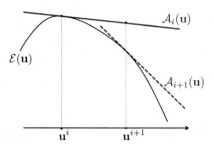

Figure 3.1 The first-order approximation of a concave function is a tight upper bound. The auxiliary function $\mathcal{A}_i(\mathbf{u})$ is depicted in red color and the original concave objective function $\mathcal{E}(\mathbf{u})$ in black. Clearly, minimizing the first-order approximation of a concave function at each iteration guarantees that the function does not increase.

Proposition 1 (Auxiliary function for a concave CRF term). *At current solution $\mathbf{u}^i$ (iteration i), we have the following auxiliary function (tight upper bound) for $\mathcal{E}(\mathbf{u})$ when Ψ is negative semi-definite:*

$$\mathcal{E}(\mathbf{u}) \leq \mathbf{u}^t \ln \mathbf{u} + \mathbf{u}^t \mathbf{v} + \frac{1}{2}(\mathbf{u}^i)^t \Psi \mathbf{u}^i + (\Psi \mathbf{u}^i)^t (\mathbf{u} - \mathbf{u}^i). \qquad (3.10)$$

Proof. Ψ is negative semi–definite means that function $\frac{1}{2}\mathbf{u}^t \Psi \mathbf{u}$ is concave and, therefore, is upper bounded by its first-order approximation at $\mathbf{u}^i$ [18]; see the illustration in Fig. 3.1. It easy to verify that this upper bound is tight, i.e., equal to $\mathcal{E}$, at current solution $\mathbf{u}^i$. $\square$

Proposition 1 means that, when we optimize iteratively the bound in Eq. (3.10), subject to the simplex constraints $\mathbf{u}_p \in \nabla_L \ \forall p$, we guarantee that the original function in Eq. (3.8) does not increase after each iteration.

Notice that the bound and the simplex constraints are separable over variables $\mathbf{u}_p$. In fact, we can express, up to a constant, the auxiliary function in the form of sum of independent functions, each corresponding to a pixel p:

$$\sum_{p \in \Omega} \mathbf{u}_p^t \ln \mathbf{u}_p + \mathbf{u}_p^t (\mathbf{v}_p + \mathbf{a}_p^i), \qquad (3.11)$$

where:

$$\mathbf{a}_p^i = (a_{p,1}^i, \dots a_{p,L}^i)^t$$
$$a_{p,l}^i = \sum_q \sum_{l'} w_{p,q} c(l, l') u_{q,l'}^i = \sum_q \sum_{l'} \psi_{p,q}(l, l') u_{q,l'}^i. \qquad (3.12)$$

Therefore, when we have a valid auxiliary function in (3.10), i.e., Ψ is negative semi-definite, we can minimize the objective with respect to each $\mathbf{u}_p$, with the latter constrained to be within the probability simplex, while guaranteeing convergence to a local minimum of $\mathcal{E}$:

$$\min_{\mathbf{u}_p \in [0,1]^L} \mathbf{u}_p^t \ln \mathbf{u}_p + \mathbf{u}_p^t(\mathbf{v}_p + \mathbf{a}_p^i) \quad \text{s.t.} \quad \mathbf{1}^t \mathbf{u}_p = 1. \tag{3.13}$$

Note that negative entropy term $\mathbf{u}_p^t \ln \mathbf{u}_p$ restricts $\mathbf{u}_p$ to be nonnegative, which removes the need for explicitly adding constraints $\mathbf{u}_p \geq 0$. This term is convex and, therefore, the problem in (3.13) is convex: The objective is convex because it is the sum of linear and convex functions, and the constraint is affine. Therefore, one can minimize this convex problem for each p by solving the Karush–Kuhn–Tucker (KKT) conditions[1]:

$$\mathbf{v}_p + \mathbf{a}_p^i + (1 + \gamma_p)\mathbf{1} + \ln \mathbf{u}_p = 0$$
$$\mathbf{1}^t \mathbf{u}_p = 1. \tag{3.14}$$

The first equation in (3.14) is obtained by converting each problem to an unconstrained one via the Lagrangian and setting the derivative with respect to $\mathbf{u}_p$ equal to zero. γ_p is the Lagrange multiplier for the simplex constraint of pixel p. The KKT conditions have a closed-form solution of each variable $\mathbf{u}_p$. It is easy to see that the first equation in (3.14) yields the following solution for variable $u_{p,l}$, i.e., the update of this variable at iteration $i + 1$:

$$u_{p,l}^{i+1} = \frac{\exp(-v_{p,l} - a_{p,l}^i)}{\exp(1 + \gamma_p)}. \tag{3.15}$$

Plugging (3.15) into the simplex constraint in (3.14) yields the following optimality conditions for dual variables (Lagrange multipliers) γ_p:

$$\exp(1 + \gamma_p) = \sum_{l'} \exp(-v_{p,l'} - a_{p,l'}^i). \tag{3.16}$$

This factor can be viewed as a normalization constant for $u_{p,l}$. Indeed, plugging (3.16) into (3.15) yields the following normalized, closed-form

---

[1] Note that strong duality holds since the objectives are convex and the simplex constraints are affine. This means that the solutions of the (KKT) conditions minimize the auxiliary function.

updates for each variable $u_{p,l}$:

$$u_{p,l}^{i+1} = \frac{\exp(-v_{p,l} - a_{p,l}^i)}{\sum_{l'} \exp(-v_{p,l'} - a_{p,l'}^i)}. \tag{3.17}$$

Notice that, within each iteration, we can compute the update of assignment variable $u_{p,l}$ in Eq. (3.15) independently of the other assignment variables. Therefore, for large $|\Omega|$, as in the case of high-resolution images or 3D grids, the updates in (3.17) can be implemented in parallel to ease the computation time. In these updates, the computation of potential $a_{p,l'}^i$ for each pixel p is $O(|\Omega|)$. This step, often referred to as message passing [1,17], requires a summation over all the pixels in the image. Therefore, the complexity of updating all the variables in (3.17) is $O(|\Omega|^2)$, which is not practical for computer vision problems. The authors of the popular work in [1] noticed that, in the case of Gaussian kernels, high-dimensional filtering[2] techniques [19] can be used to update all the variables concurrently in $O(|\Omega|)$ time. This reduces the quadratic complexity of a naive implementation to linear, yielding very fast mean-field algorithms in practice [1,2].

3.2.3 The general case of a nonconcave pairwise term

The convergence of the parallel updates in Eq. (3.17) to a local minimum of $\mathcal{E}$ is guaranteed only when matrix Ψ is negative semi-definite, i.e., the upper bound in (3.10) is a valid auxiliary function. This is the case, for instance, when using Gaussian kernels for pairwise potentials $w_{p,q}$ and the Potts label-compatibility matrix. More generally, there are two common conditions on pairwise-potential matrix $W = [w_{p,q}]$ and label-compatibility matrix $C = [c(l, l')]$, guaranteeing that parallel message passing converges: (1) W is positive semi-definite and (2) there is a constant h for which $C' = [c(l, l') + h]$ is negative semi-definite. Notice that adding a constant to all the entries of the label compatibility matrix C does not change the problem of minimizing the KL divergence in Eq. (3.7). It just adds a constant[3] independent of $\mathbf{u}$ to the function, resulting in an equivalent problem. For example, in the case of Potts, all the diagonal elements of C are equal to zero and off-diagonal elements to 1. Therefore, using $h = -1$ results in a negative semi-definite diagonal matrix, with all the diagonal elements equal to -1.

---

[2] Fast high-dimensional filtering techniques using the permutohedral lattices are common in signal processing.

[3] The constant is $h \sum_{p,q} w_{p,q}$.

Condition (1) applies to a general class of positive semi-definite kernels, e.g., Gaussian kernels. Therefore, common models that use Potts in conjunction with Gaussian kernels [1,2] satisfy conditions (1) and (2). These models follow up to an additive constant independent of $\mathbf{u}$, the general form in Eq. (3.8), with a negative semi-definite matrix Ψ. This is due to the fact that Kronecker product $\Psi = W \otimes C$ is negative semi-definite when W is positive semi-definite (condition 1) and C is negative semi-definite (condition 2).

Proposition 2 (Generalization of the auxiliary function to nonconcave CRF terms). *For any symmetric matrix Ψ and any $\mathbf{u}$ in the L-dimensional simplex ∇_L, we have the following auxiliary function for $\mathcal{E}(\mathbf{u})$ at current solution $\mathbf{u}^i$:*

$$\mathcal{E}(\mathbf{u}) \leq \mathbf{u}^t \ln \mathbf{u} + \mathbf{u}^t \mathbf{v} + \frac{1}{2}(\mathbf{u}^i)^t \Psi \mathbf{u}^i + (\Psi \mathbf{u}^i)^t (\mathbf{u} - \mathbf{u}^i) + \beta \sum_p \mathrm{KL}(\mathbf{u}_p || \mathbf{u}_p^i) \quad (3.18)$$

for all $\beta \geq 2\, max(\alpha_{max}(\Psi), 0)$, with $\alpha_{max}(\Psi)$ the largest eigenvalue of Ψ.

In Proposition 2, KL denotes the Kullback–Leibler divergence given by:

$$\mathrm{KL}(\mathbf{u}_p || \mathbf{u}_p^i) = \sum_l u_{p,l} \ln \frac{u_{p,l}}{u_{p,l}^i} = (\mathbf{u}_p)^t \log \mathbf{u}_p - (\mathbf{u}_p)^t \log \mathbf{u}_p^i.$$

Notice that $\alpha_{max}(\Psi) \leq 0$ when Ψ is negative semi-definite, i.e., $\mathcal{E}$ is concave. In this case, the bound in (3.18) is valid for any $\beta \geq 0$. The particular case $\beta = 0$ corresponds to the upper bound in Proposition 1, in which we used the fact that a concave function is upper bounded by its first-order approximation at current solution $\mathbf{u}^i$.

Proof. For a nonconcave pairwise CRF term $\frac{1}{2}\mathbf{u}^t \Psi \mathbf{u}$, matrix Ψ has at least one strictly positive eigenvalue. Let us first consider the case of a convex pairwise term, i.e., all the eigenvalues of Ψ are nonnegative or, equivalently, Ψ is positive semi-definite. In this case, the pairwise CRF term is α-smooth for $\alpha \geq \alpha_{max}(\Psi) > 0$, with $\alpha_{max}(\Psi)$ the maximum eigenvalue of Ψ; see the details in the appendix. This is due to the fact that Ψ is the Hessian of the pairwise CRF term. As the latter is a convex, α-smooth function, we have the following upper bound on it $\forall \alpha \geq \alpha_{max}(\Psi)$; see the lemma in the appendix:

$$\frac{1}{2}\mathbf{u}^t \Psi \mathbf{u} \leq \frac{1}{2}(\mathbf{u}^i)^t \Psi \mathbf{u}^i + (\Psi \mathbf{u}^i)^t (\mathbf{u} - \mathbf{u}^i) + \alpha \|\mathbf{u} - \mathbf{u}^i\|^2$$

$$= \frac{1}{2}(\mathbf{u}^i)^t \Psi \mathbf{u}^i + (\Psi \mathbf{u}^i)^t(\mathbf{u} - \mathbf{u}^i) + \alpha \sum_p \|\mathbf{u}_p - \mathbf{u}_p^i\|^2. \qquad (3.19)$$

Using Pinsker inequality [20] and given the fact that both $\mathbf{u}_p$ and $\mathbf{u}_p^i$ are simplex vectors, we also have the following upper bound on the Euclidean distances in the last term of the upper bound in (3.19):

$$\|\mathbf{u}_p - \mathbf{u}_p^i\|^2 \leq 2\,\mathrm{KL}(\mathbf{u}_p\|\mathbf{u}_p^i). \qquad (3.20)$$

This gives the following upper bound on the pairwise CRF term $\forall \beta \geq 2\alpha \geq 2\alpha_{max}(\Psi)$:

$$\frac{1}{2}\mathbf{u}^t \Psi \mathbf{u} \leq \frac{1}{2}(\mathbf{u}^i)^t \Psi \mathbf{u}^i + (\Psi \mathbf{u}^i)^t(\mathbf{u} - \mathbf{u}^i) + \beta \sum_p \mathrm{KL}(\mathbf{u}_p\|\mathbf{u}_p^i). \qquad (3.21)$$

The result in Proposition 2 follows directly from the upper bound in (3.21).

Now, let consider the more general when Ψ is indefinite, i.e., the pairwise CRF term is neither concave, nor convex. In this case, matrix Ψ has both strictly positive and strictly negative eigenvalues. Because Ψ is a real symmetric matrix, we can express it using its eigenvalue decomposition: $\Psi = P^t D P$. $D = \mathrm{diag}(\lambda_j, 1 \leq j \leq L|\Omega|)$ is a diagonal matrix whose entries are the eigenvalues of Ψ, and P is an orthogonal matrix whose rows are the eigenvectors of Ψ. It is easy to see that D can be expressed as $D = D_1 + D_2$, where $D_1 = \mathrm{diag}(\lambda_j, 1 \leq j \leq d^+)$ is the diagonal matrix, whose diagonal contains only the positive eigenvalues of Ψ, with d^+ the number of positive eigenvalues, and $D_2 = \mathrm{diag}(\lambda_j, d^+ + 1 \leq j \leq L|\Omega|)$ the diagonal matrix including only the negative eigenvalues of Ψ. Clearly, Ψ can be written as $\Psi_1 + \Psi_2$, with $\Psi_1 = P^t D_1 P$ positive semi-definite and $\Psi_2 = P^t D_2 P$ negative semi-definite. Therefore, the pairwise CRF term can be written as the sum of a concave and a convex function $(\mathbf{u})^t \Psi \mathbf{u} = (\mathbf{u})^t \Psi_1 \mathbf{u} + (\mathbf{u})^t \Psi_1 \mathbf{u}$. Using the upper bound in Eq. (3.21) for the convex part and the first-order approximation for the concave part, as in Proposition 1, we obtain the result for the general case. □

It is worth noting that the quadratic upper bound in Eq. (3.19) is a valid auxiliary function, and, in principle, we could have used it as well. Introducing the KL bound in (3.20) to replace quadratic terms $\|\mathbf{u}_p - \mathbf{u}_p^i\|^2$ yields a different auxiliary function, but seems unwarranted at the first look. However, these KL terms have important computational advantages over quadratic terms in the case of simplex constraints, more so when the number of points is very large, as is the case of segmentation problems. In fact, with the KL terms, at each iteration and for each pixel p, the KKT conditions have closed-form solutions for variables $\mathbf{u}_p$, yielding a generalization of (3.17):

$$u_{p,l}^{i+1} = \frac{(u_{p,l}^i)^{\frac{\beta}{\beta+1}} \exp\left(\frac{-v_{p,l}-d_{p,l}^i}{\beta+1}\right)}{\sum_{l'}(u_{p,l'}^i)^{\frac{\beta}{\beta+1}} \exp\left(\frac{-v_{p,l'}-d_{p,l'}^i}{\beta+1}\right)}. \tag{3.22}$$

With quadratic terms $\|\mathbf{u}_p - \mathbf{u}^i_p\|^2$, this is not the case that we would need to solve iteratively the KKT conditions with respect to each variable $\mathbf{u}_p$. The updates in Eq. (3.22) were recently introduced in [11]. Notice that, for $\beta = 0$, the updates in (3.22) reduce to the softmax updates we obtained earlier in Eq. (3.17) for the case of a concave CRF term. Also, similarly to (3.17), the more general updates in (3.22) could be performed independently for each pixel at each iteration, while guaranteeing convergence, yielding a scalable, computationally efficient optimizer for segmentation problems.

Appendix 3.A

Definition 1 (Lipschitz gradient). A convex function $\mathcal{E}$ is α-smooth if the gradient of $\mathcal{E}$ is Lipschitz, with $\alpha > 0$ being a Lipschitz constant:

$$\|\nabla\mathcal{E}(\mathbf{f}) - \nabla\mathcal{E}(\mathbf{g})\| \leq \alpha \|\mathbf{f} - \mathbf{g}\| \quad \forall \mathbf{f}, \mathbf{g}.$$

Equivalently, there exists a strictly positive β such that the Hessian of $\mathcal{E}$ verifies

$$\nabla^2\mathcal{E} \preceq \alpha\mathbf{I},$$

where $\mathbf{I}$ denotes the identity matrix.

Remark 1. Let $\alpha_{max}(\nabla^2\mathcal{E})$ denotes the maximum eigenvalue of the Hessian of function $\mathcal{E}$, and assume $\alpha_{max}(\nabla^2\mathcal{E})$ is strictly positive. In this case, $\alpha_{max}(\nabla^2\mathcal{E})$ is a valid Lipschitz constant for the gradient of $\mathcal{E}$ because

$$\nabla^2\mathcal{E} \preceq \alpha_{max}(\nabla^2\mathcal{E})\,\mathbf{I}.$$

Lemma 1 (Upper bound on a α-smooth function). *If a convex function $\mathcal{E}$ is α-smooth, then we have the following upper bound on $\mathcal{E}$:*

$$\mathcal{E}(\mathbf{f}) \leq \mathcal{E}(\mathbf{g}) + \nabla\mathcal{E}(\mathbf{g})^t(\mathbf{f} - \mathbf{g}) + \alpha\|\mathbf{f} - \mathbf{g}\|^2. \tag{3.23}$$

Proof. The proof of inequality (3.23) is straightforward. It suffices to start from convexity condition $\mathcal{E}(\mathbf{f}) \leq \mathcal{E}(\mathbf{g}) + \nabla\mathcal{E}(\mathbf{f})^t(\mathbf{f} - \mathbf{g})$, and to use the Cauchy–Schwarz inequality and Lipschitz-gradient condition for a α-smooth function:

$$\mathcal{E}(\mathbf{f}) \leq \mathcal{E}(\mathbf{g}) + \nabla\mathcal{E}(\mathbf{f})^t(\mathbf{f} - \mathbf{g})$$
$$= \mathcal{E}(\mathbf{g}) + \nabla\mathcal{E}(\mathbf{g})^t(\mathbf{f} - \mathbf{g}) + \left(\nabla\mathcal{E}(\mathbf{f}) - \nabla\mathcal{E}(\mathbf{g})\right)^t(\mathbf{f} - \mathbf{g})$$
$$\leq \mathcal{E}(\mathbf{g}) + \nabla\mathcal{E}(\mathbf{g})^t(\mathbf{f} - \mathbf{g}) + \|\nabla\mathcal{E}(\mathbf{f}) - \nabla\mathcal{E}(\mathbf{g})\| \, \|\mathbf{f} - \mathbf{g}\|$$
$$\leq \mathcal{E}(\mathbf{g}) + \nabla\mathcal{E}(\mathbf{g})^t(\mathbf{f} - \mathbf{g}) + \alpha \, \|\mathbf{f} - \mathbf{g}\|^2. \qquad \square$$

References

[1] P. Krähenbühl, V. Koltun, Efficient inference in fully connected CRFs with Gaussian edge potentials, in: Advances in Neural Information Processing Systems (NIPS), 2011, pp. 109–117.

[2] A. Arnab, S. Zheng, S. Jayasumana, B. Romera-Paredes, M. Larsson, A. Kirillov, et al., Conditional random fields meet deep neural networks for semantic segmentation: Combining probabilistic graphical models with deep learning for structured prediction, IEEE Signal Processing Magazine 35 (1) (2018) 37–52.

[3] Y. Boykov, M.P. Jolly, Interactive graph cuts for optimal boundary and region segmentation of objects in n-d images, in: IEEE International Conference on Computer Vision (ICCV), 2001, pp. 105–112.

[4] Y. Boykov, G. Funka Lea, Graph cuts and efficient n-d image segmentation, International Journal of Computer Vision 70 (2) (2006) 109–131.

[5] L.C. Chen, G. Papandreou, I. Kokkinos, K. Murphy, A.L. Yuille, Semantic image segmentation with deep convolutional nets and fully connected CRFs, in: International Conference on Learning Representations (ICLR), 2015, pp. 1–14.

[6] L. Chen, G. Papandreou, I. Kokkinos, K. Murphy, A.L. Yuille, DeepLab: Semantic image segmentation with deep convolutional nets, atrous convolution, and fully connected CRFs, IEEE Transactions on Pattern Analysis and Machine Intelligence 40 (4) (2018) 834–848.

[7] M. Li, A. Shekhovtsov, D. Huber, Complexity of discrete energy minimization problems, in: European Conference on Computer Vision (ECCV), Part II, 2016, pp. 834–852.

[8] Y. Boykov, V. Kolmogorov, An experimental comparison of min-cut/max-flow algorithms for energy minimization in vision, IEEE Transactions on Pattern Analysis and Machine Intelligence 26 (9) (2004) 1124–1137.

[9] Y. Boykov, O. Veksler, R. Zabih, Fast approximate energy minimization via graph cuts, IEEE Transactions on Pattern Analysis and Machine Intelligence 23 (11) (2001) 1222–1239.

[10] V. Vineet, J. Warrell, P.H.S. Torr, Filter-based mean-field inference for random fields with higher-order terms and product label-spaces, International Journal of Computer Vision 110 (3) (2014) 290–307.

[11] P. Baqué, T.M. Bagautdinov, F. Fleuret, P. Fua, Principled parallel mean-field inference for discrete random fields, in: IEEE Conference on Computer Vision and Pattern Recognition (CVPR), 2016, pp. 5848–5857.

[12] M. Rajchl, M.C.H. Lee, O. Oktay, K. Kamnitsas, J. Passerat-Palmbach, W. Bai, et al., Deepcut: Object segmentation from bounding box annotations using convolutional neural networks, IEEE Transactions on Medical Imaging 36 (2) (2017) 674–683.

[13] A. Blake, P. Kohli, C. Rother, Markov Random Fields for Vision and Image Processing, MIT Press, 2011.

[14] M. Saito, T. Okatani, K. Deguchi, Application of the mean field methods to MRF optimization in computer vision, in: IEEE Conference on Computer Vision and Pattern Recognition (CVPR), 2012, pp. 1680–1687.

[15] S. Zheng, S. Jayasumana, B. Romera-Paredes, V. Vineet, Z. Su, D. Du, et al., Conditional random fields as recurrent neural networks, in: IEEE International Conference on Computer Vision (ICCV), 2015, pp. 1529–1537.

[16] Z. Liu, X. Li, P. Luo, C.C. Loy, X. Tang, Deep learning Markov random field for semantic segmentation, IEEE Transactions on Pattern Analysis and Machine Intelligence 40 (8) (2018) 1814–1828.

[17] P. Krähenbühl, V. Koltun, Parameter learning and convergent inference for dense random fields, in: International Conference on Machine Learning (ICML), 2013, pp. 513–521.

[18] K. Lange, D.R. Hunter, I. Yang, Optimization transfer using surrogate objective functions, Journal of Computational and Graphical Statistics 9 (1) (2000) 1–20.

[19] A. Adams, J. Baek, M.A. Davis, Fast high-dimensional filtering using the permutohedral lattice, Computer Graphics Forum 29 (2) (2010) 753–762.

[20] I. Csiszar, J. Korner, Information Theory: Coding Theorems for Discrete Memoryless Systems, 2nd ed., Cambridge University Press, 2011.

CHAPTER 4

Regularized model fitting

4.1 General probabilistic form

In general, image segmentation models integrate regularization terms with probabilistic models that take the form of negative log-likelihoods. For instance, a log-likelihood term typically uses parametric statistical models of the observed features (e.g., intensity or colors), with each model fitting the data within one of the segmentation regions via a set of parameters. In Chapter 1, we discussed the popular Boykov–Jolly functional for segmentation [5,6], which can be written as follows in the case of L regions:

$$
\mathcal{E}(\mathbf{S}) = -\sum_{l=0}^{L-1} \sum_{p \in S^l} \ln \Pr(\mathbf{f}_p | l, \boldsymbol{\theta}) + \lambda \mathcal{R}(\mathbf{S}), \tag{4.1}
$$

where:

- $\mathbf{S} = \{S^l, l = 0, \dots, L-1\}$ denotes a variable partition of image domain $\Omega \subset \mathbb{R}^M$, which is a set of pixels ($M = 2$), voxels ($M = 3$) or other (higher-dimensional) spatial locations p.
- $\Pr(\mathbf{f}_p | l, \boldsymbol{\theta})$ is a given probability model (or distribution) of image data within segmentation region S^l, $l = 0, \dots, L-1$. $\boldsymbol{\theta}$ is a set of parameters characterizing these distributions, and $\mathbf{f}_p \in \mathbb{R}^N$ denotes the observed feature at spatial location p. These features might be intensity values in $\mathbb{R}^1$, colors in $\mathbb{R}^3$ or some other features in higher-dimensional spaces.

The first term in (4.1) penalizes the deviation of image features within region S^l from likelihood model $\Pr(. | l, \boldsymbol{\theta})$. The second, $\mathcal{R}$, is a regularization term evaluating segmentation boundary length via Euclidean or some edge-sensitive, feature-weighted metric. In the BJ model [5,6], likelihood probabilities $\Pr(. | l, \boldsymbol{\theta})$ are fixed. They are learned *a priori,* either from user inputs or from training images. In this case, the first term in (4.1) is a sum of unary (first-order) potentials, and the second is pairwise submodular. Therefore, the global optimum of the overall functional can be obtained with powerful discrete combinatorial optimization techniques via a single graph cut [7].

High-Order Models in Semantic Image Segmentation
https://doi.org/10.1016/B978-0-12-805320-1.00009-9

When models $\Pr(.|l, \boldsymbol{\theta})$ are not known, it is very common to proceed in an unsupervised fashion, assuming parameters $\boldsymbol{\theta}$ are now continuous variables of the model:

$$\mathcal{J}(\mathbf{S}, \boldsymbol{\theta}) = -\sum_{l=0}^{L-1} \sum_{p \in S^l} \ln \Pr(\mathbf{f}_p | l, \boldsymbol{\theta}) + \lambda \mathcal{R}(\mathbf{S}). \tag{4.2}$$

The objective in (4.2) is now mixed, depending on two types of variables, discrete segmentation variables $\mathbf{S} = \{S^l, l = 0, \dots, L - 1\}$ and continuous distribution parameters $\boldsymbol{\theta} = (\boldsymbol{\theta}_0, \dots, \boldsymbol{\theta}_{L-1})$, with $\boldsymbol{\theta}_l$ denoting the set of parameters of region S^l. In fact, there is a dependence now between $\boldsymbol{\theta}_l$ and $\mathbf{S}_l$, typically nonlinear that makes the ensuing optimization problem much more difficult than minimizing (4.1). With such a dependence, the likelihood term becomes high-order; it is not the sum of unary potentials anymore. As we will see later, for some forms of $\Pr(.|l, \boldsymbol{\theta})$, one can express the likelihoods solely in terms of segmentation variables $\mathbf{S}$ by replacing the parameters with the corresponding optimal, region-dependent values.

When the parameters are variable, the first term (4.2) can be seen as a *maximum likelihood* (ML) model-fitting cost, which has a straightforward connection to the standard K-means clustering objective and its probabilistic variants [3]. In fact, image segmentation can be viewed as data clustering subject to spatial regularization constraints. The term clustering is widely used in the general context of data partitioning [8], in a large spectrum of areas other than computer vision. Without regularization, a large class of popular segmentation objectives [1,2,9,10] reduce to clustering functionals that view image features $\mathbf{f}_p$ as arbitrary data points indexed by p. In segmentation, however, data points $\mathbf{f}_p \in \mathbb{R}^N$ (e.g., intensity, colors or textures) are sampled at regularly spaced image locations (or pixels) $p \in \mathbb{R}^M$. Such spatial locations are crucial, and motivate a breadth of regularization methods in computer vision.

For mixed functional (4.2), it is very common to follow an iterative two-step procedure [1,2,9,10], one step fixing each segment S^l and minimizing $\mathcal{J}$ with respect to parameters $\boldsymbol{\theta}_l$ and the other finding the optimal segmentation with the parameters fixed. This is the case of several popular segmentation algorithms such as the GrabCut algorithm of Rother et al. [1], the Chan–Vese model [2], the region competition algorithm of Zhu and Yuille [9], among others [10].

4.2 Standard models

4.2.1 K-means and the Chan–Vese model

The link between the well-known K-means clustering objective [11,12] and the general likelihood term in (4.2) is straightforward. Let us assume that the features are one-dimensional ($\mathbf{f}_p \in \mathbb{R}$), and follow a Gaussian distribution within each of the segmentation regions:

$$\Pr(\mathbf{f}_p | l, \boldsymbol{\theta}) = \frac{1}{\sqrt{2\pi\sigma^2}} \exp\left(-\frac{\|\mathbf{f}_p - \mathbf{m}^l\|^2}{2\sigma^2}\right). \tag{4.3}$$

In Eq. (4.3), we further assumed that all the regions have the same fixed standard deviation σ. Each region S^l is, therefore, characterized by its mean parameter $\mathbf{m}^l$. $\boldsymbol{\theta} = (\mathbf{m}_0, \ldots, \mathbf{m}_{L-1})$ denotes the set of all mean parameters, and $\|.\|$ is the Euclidean metric. By plugging Gaussian models (4.3) into (4.2), it is easy to see that the general log-likelihood term is equivalent, up to additive and multiplicative constants, to the following objective:

$$\sum_l \sum_{p \in S^l} \|\mathbf{f}_p - \mathbf{m}^l\|^2. \tag{4.4}$$

This functional, which corresponds exactly to the K-means objective [11,12], has two types of variables, a set of discrete variables describing segmentation (or clustering) and another set of continuous variables encoding region parameters.

Extension of the link between K-means objective (4.4) and likelihood model (4.2) to vector-valued features $\mathbf{f}_p$ is straightforward. Considering a vector-valued Gaussian distribution, the log-likelihoods can be expressed as follows:

$$\begin{aligned}-\ln \Pr(\mathbf{f}_p | l, \boldsymbol{\theta}) &\overset{c}{=} \mathcal{G}^{vect}(\mathbf{f}_p, \mathbf{m}^l, \boldsymbol{\Sigma}^l) \\ &= \frac{1}{2}\ln(\det \boldsymbol{\Sigma}^l) + \frac{1}{2}(\mathbf{f}_p - \mathbf{m}^l)^t (\boldsymbol{\Sigma}^l)^{-1}(\mathbf{f}_p - \mathbf{m}^l).\end{aligned} \tag{4.5}$$

Symbol $\overset{c}{=}$ denotes equality up to an additive constant, and $\boldsymbol{\Sigma}^l$ is the covariance matrix of the features within region $\mathbf{S}^l$. Choosing $\boldsymbol{\Sigma}^l = \sigma \mathbf{I} \,\forall l$, with $\mathbf{I}$ the identity matrix, it easy to see that plugging (4.5) into the likelihood term yields the K-means objective, up to additive and multiplicative constants.

For fixed segments, optimizing (4.4) with respect to each parameter $\mathbf{m}^l$ gives a closed-form solution[1] that turns out to be the mean of the observed features within the corresponding region:

$$\tilde{\mathbf{m}}^l = \frac{1}{|S^l|} \sum_{p \in S^l} \mathbf{f}_p. \tag{4.6}$$

|.| denotes the number of pixels (cardinality) when its argument is a region. Starting from an initial partition, the standard K-means procedure alternates two steps until convergence, one step updates the mean of features within each segment S^l with the partition fixed, and the other updates the region membership of each point p with the parameters fixed:

1. Initialize partition $\mathbf{S}$
2. Iterate until convergence:
    a. Fix $\mathbf{S}$ and update the mean of each cluster (region) according to (4.6)
    b. Fix $\mathbf{m}$ and update the membership each p as follows:

$$x_p = \arg\min_l \|\mathbf{f}_p - \mathbf{m}^l\|, \tag{4.7}$$

where x_p denotes a cluster assignment variable for p: $S^l = \{p | x_p = l\}$. The standard updates the guarantee that the K-means function decreases at each iteration, and converge to a local minimum of cost function (4.4).

It is well known that the K-means problem is NP hard [13]. In fact, it is possible to express such an objective as a high-order function that depends only on partition variable $\mathbf{S}$. Embedding the optimal parameters in (4.6) into objective function (4.4) yields the following:

$$\sum_l \sum_{p \in S^l} \|\mathbf{f}_p - \frac{1}{|S^l|} \sum_{p \in S^l} \mathbf{f}_p\|^2. \tag{4.8}$$

Expanding the Euclidean distances, and ignoring a constant independent of the partition, we obtain the following equivalent ratio functional, which depends on the inner products between pairs of features [13]:

$$\sum_l \frac{1}{|S^l|} \sum_{p,q \in S^l} \mathbf{f}_p^t \mathbf{f}_q. \tag{4.9}$$

---

[1] For each sum over segment S^l, we set the derivative of the sum with respect to $\mathbf{m}^l$ equal to zero, and compute the corresponding closed-form solution.

In fact, not only K-means but all the likelihood models that we will be discussing in this chapter are high-order terms, due to the dependence between the parameters and segmentation variables. The popular Chan–Vese functional [2] is a basic example of the general model fitting and regularization objective in (4.2). It combines the K-means term with a continuous level-set based length regularization. The algorithm alternates mean computations, as in the standard K-means procedure, and curve evolution updates implemented via level sets, which replace the basic region updates (4.7) in K-means. The curve evolution is a partial differential equation derived from gradient-descent optimization of the functional [2]. The equation contains two velocities defined at each point p. The first is data-driven and has an effect similar to (4.7): It encourages point p to belong to the region having the closest mean to feature $\mathbf{f}_p$ (according to the Euclidean distance). The second is a curvature velocity that regularizes the segmentation boundaries.

4.2.2 Other parametric distributions

A straightforward way to generalize the Chan–Vese objective is to consider both the means and covariance matrices of the Gaussian distributions in (4.2) as variables [14]. With such an extension, the model parameters of (4.2) become $\boldsymbol{\theta} = \{\mathbf{m}^l, \boldsymbol{\Sigma}^l, l = 0, \dots, L-1\}$, and are not anymore in the same space as feature points $\mathbf{f}_p$.

The classical regularized K-means objective [2,15], which assumes the features within each region follow a piecewise constant model and its Gaussian generalization [14], also referred to as elliptic K-means [13], have been widely in segmenting intensity images acquired by conventional cameras. This is mainly due to the fact that such Gaussian descriptions result in simple, closed-form updates of region parameters, i.e., sample means and variances. Although practical in many scenarios, such Gaussian models may not be adequate to handle various applications and problems. For instance, image features acquired by sensors other than standard cameras are, typically, not Gaussian distributed. This is the case in radar [16,17], sonar [18] and medical images [19]. For instance, synthetic aperture radar (SAR) images, which are common in remote sensing, follow the gamma/exponential model [17]. In some applications, each segmentation region may need a model different from the other regions in the image. For instance, in sonar imagery [18], the luminance within shadow regions follow the Gaussian distribution, whereas Rayleigh is more appropriate in the reverberation regions. It is possible to use a model that can adapt to different image distributions, e.g., the Weibull distribution [20]. In this case, however, optimization with respect to region parameters does not yield closed-form solutions anymore, which may increase substantially the computational cost of the ensuing algorithms.

4.2.3 General distortions and K-modes

It is possible to generalize K-means by using other distortion measures $\|.\|_d$ instead of the squared Euclidean distances in (4.4). This yields the following general objective function, which has been used in the general context of clustering [4,21], and integrated with spatial regularization in the context of image segmentation [22,23]:

$$\sum_l \sum_{p \in S^l} \|\mathbf{f}_p - \mathbf{m}^l\|_d. \tag{4.10}$$

Notice that we are still able to express (4.10) in the form of a log-likelihood term:

$$\|\mathbf{f}_p - \mathbf{m}^l\|_d = -\ln \Pr(\mathbf{f}_p|l, \mathbf{m}) = -\ln\left(\frac{1}{Z_d} \exp(-\|\mathbf{f}_p - \mathbf{m}^l\|_d)\right), \tag{4.11}$$

where $\Pr(.|l, \mathbf{m})$ takes the form of a Gibbs model and Z_d is a normalization constant. For this general case, optimal parameter $\mathbf{m}^l$ does not necessarily correspond to the mean of features within region S^l. In the following, we will examine kernel-induced, non-Euclidean distances between the features and the parameters. In this case, optimization with respect to parameter $\mathbf{m}^l$ yields the mode of the density of features within region (or cluster) S^l, and the objective is referred to as K-modes [4,21]. A mode of a region (or cluster) S^l is the maximum of the density of features within S^l. Therefore, segmentation methods that use kernel-induced distances [22,23] do not fit a statistical image distribution of specific shape such as Gaussian or Rayleigh. Instead, we find the mode of the image density within region S^l and assign pixels (or data points) to such a mode via a kernel-induced distortion measure. This can be very convenient in practice because we do not need to know *a priori* the correct image model within each segment. For instance, the work in [23] showed that, in practice, a regularized K-modes objective can deal with images of various distributions, obtaining consistently similar or higher performances than choosing the correct statistical model (e.g., Gamma, exponential or Gaussian). Furthermore, in practical scenarios where image distributions vary within the image, K-modes affords more

Table 4.1 Examples of prevalent kernel functions $K(\mathbf{y}, \mathbf{z})$.

RBF kernel	Polynomial kernel	Sigmoid kernel
$\exp(-\|\mathbf{y} - \mathbf{z}\|^2/\alpha^2)$	$(\mathbf{y}^t\mathbf{z} + \beta)^\alpha$	$\tanh(\alpha(\mathbf{y} \cdot \mathbf{z}) + \beta)$

versatility in comparison to choosing a single statistical model. For instance, synthetic-aperture-radar distribution within a single image can vary from Gaussian to Gamma [24].

4.2.4 Kernel-induced distances

Let $\phi(.)$ denotes a mapping from the initial space of data points (e.g., colors) to a higher, possibly infinite-dimensional feature space. Regularized K-modes minimizes the following objective function:

$$\sum_{l=0}^{L-1} \sum_{p \in S^l} \|\phi(\mathbf{f}_p) - \phi(\mathbf{m}^l)\| + \lambda \mathcal{R}(\mathbf{S}). \tag{4.12}$$

The data term in (4.12) evaluates kernel-induced (non-Euclidean) distances between the features and parameters. In fact, we do not need to know explicitly mapping ϕ. We can express any symmetric, positive semi-definite kernel function as an Euclidean dot product in some high-dimensional space, as stated by the Mercer's theorem:

$$K(\mathbf{y}, \mathbf{z}) = \phi(\mathbf{y})^t \phi(\mathbf{z}). \tag{4.13}$$

Table 4.1 lists several common kernel functions.

Using the kernel function in the distortions measures of (4.12) yields the following:

$$\begin{aligned}
\|\phi(\mathbf{f}_p) - \phi(\mathbf{m}^l)\|^2 &= (\phi(\mathbf{f}_p) - \phi(\mathbf{m}^l))^t(\phi(\mathbf{f}_p) - \phi(\mathbf{m}^l)) \\
&= \phi(\mathbf{f}_p)^t\phi(\mathbf{f}_p) + \phi(\mathbf{m}_l)^t\phi(\mathbf{m}_l) - 2\phi(\mathbf{f}_p)^t\phi(\mathbf{m}^l) \\
&= K(\mathbf{f}_p, \mathbf{f}_p) + K(\mathbf{m}^l, \mathbf{m}^l) - 2K(\mathbf{f}_p, \mathbf{m}^l).
\end{aligned} \tag{4.14}$$

Notice that, for kernels $K(\mathbf{y}, \mathbf{z})$ that depend only on the difference $(\mathbf{y} - \mathbf{z})$, the first and second kernels in the last line of (4.14) add a constant to objective function (4.12). This is the case, for instance, of the commonly used Radial Basis Function (RBF) kernel; see Table 4.1. Considering (4.14) for this class of translation-invariant kernels, and ignoring the constant, objective function (4.12) becomes equivalent to:

$$-\sum_{l=0}^{L-1}\sum_{p\in S^l}K(\mathbf{f}_p,\mathbf{m}^l)+\lambda\mathcal{R}(\mathbf{S}). \tag{4.15}$$

To see the link between K-modes and region (or cluster) density, let us first consider the following definition that provides an estimator of the distribution of features within a given set of discrete points.

Definition 1 (Kernel Density Estimation). Given a set of points $S \subset \Omega$ and a set of features $\mathbf{f}_p \in \mathbb{R}^N$ defined for each point $p \in S$, the kernel density estimate (KDE) of the distribution of features $\mathbf{f}_p$ within set S is given by:

$$P_S(\mathbf{z}) = \frac{1}{|S|}\sum_{p\in S}K(\mathbf{z},\mathbf{f}_p). \tag{4.16}$$

Now, using (4.16), it is easy to see that objective (4.15) can be written as follows:

$$-\sum_{l=0}^{L-1}|S^l|P_{S^l}(\mathbf{m}^l)+\lambda\mathcal{R}(\mathbf{S}). \tag{4.17}$$

From this expression, it becomes clear that, when the regions are fixed, optimal parameter $\tilde{\mathbf{m}}^l$ is the mode (i.e., the data point with the highest density) of the distribution of features within region S^l.

Regularized K-modes (4.15) can be optimized by alternating two steps, one seeking the density mode for each cluster with the segments fixed and the other fixes the modes and updates the segmentation, e.g., via graph cuts [23]. There are several possible options for finding the modes. For instance, one can search exhaustively for the feature point having the highest density within the region, albeit this choice might be expensive computationally. A standard solution would be to compute the necessary conditions for a minimum of the objective function with respect to the parameters, which, in the case of the RBF kernel, are [22]:

$$\mathbf{m}_l - f(S^l,\mathbf{m}^l) = 0, \quad l \in [0,\dots,L-1], \tag{4.18}$$

where function f_S is given by

$$f(S^l,\mathbf{m}^l) = \frac{\sum_{p\in S^l}K(\mathbf{f}_p,\mathbf{m}^l)\mathbf{f}_p}{\sum_{p\in S^l}K(\mathbf{f}_p,\mathbf{m}^l)}, \quad l \in [0,\dots,L-1]. \tag{4.19}$$

Fixed-point iterations that consist of the following updates yield a minimum with respect to the parameters:

$$\mathbf{m}_{t+1}^l = f(S^l, \mathbf{m}_t^l), \; l \in [0, \ldots, L-1], \; t = 1, 2, \ldots, \qquad (4.20)$$

where t denotes the iteration number. The updates in (4.20) can be viewed as kernel-weighted means, which take the form of the very popular mean-shift iterations for mode seeking and clustering [25,26]. For kernels of the form $K(\mathbf{y}, \mathbf{z}) = cg(\|\mathbf{y} - \mathbf{z}\|)$, with g a convex and monotonically decreasing function,[2] these updates are guaranteed to converge to a mode and yield monotonic increases in the corresponding density estimates: $P_{S^l}(\mathbf{m}_{t+1}^l) \geq P_{S^l}(\mathbf{m}_t^l)$; see Theorem 1 in [25]. Notice that the widely used RBF belongs to this form of kernels. At convergence, the limit of the sequence of updates in (4.20), $\tilde{\mathbf{m}}^l$, is a fixed point of function $f(S^l, .)$, i.e., it verifies: $f(S^l, \tilde{\mathbf{m}}^l) = \tilde{\mathbf{m}}^l$. This can be seen from

$$\tilde{\mathbf{m}}^l = \lim_{t \to +\infty} \mathbf{m}_{t+1}^l = \lim_{t \to +\infty} f(S^l, \mathbf{m}_t^l) = f(S^l, \lim_{t \to +\infty} \mathbf{m}_t^l) = f(S^l, \tilde{\mathbf{m}}^l). \quad (4.21)$$

Therefore, $\tilde{\mathbf{m}}^l$ is a solution for the necessary condition in (4.18). The procedure optimizing (4.15) can be summarized as follows:
1. Initialize partition **S**
2. Iterate until convergence:
    a. Iterate until convergence:
        – For each l, Fix S^l and find mode $\tilde{\mathbf{m}}^l$ by mean-shift iterations (4.20)
    b. Fix $\mathbf{m}^l \; \forall l$ and solve (4.15) via graph cuts [23] or other optimization techniques [22].

Mean-shift updates are standard in the context of feature-space analysis and clustering [25]. Such iterations have been traditionally used to find the stationary points of the density of features. Starting from a given feature point, the updates converge to a mode. It is interesting to see that similar updates appear in the context of optimizing the K-modes cost function. In that sense, K-modes can be interpreted as function-minimization formulation of standard mean-shift procedures. Of course, this does not mean that they are equivalent; K-modes function assumes the number of clusters is known, which is not the case of standard mean-shift procedures for density analysis [25].

---

[2] Strictly positive constant c ensures that K integrates to one.

Figure 4.1 GrabCut segmentation [1]. Interactive foreground/background segmentation of color images from a bounding box specified by the user input. GrabCut characterizes the color features within the regions by highly descriptive models such as GMMs. (*Figure from [1] © ACM.*)

4.2.5 GrabCut and highly descriptive models

Rother et al. [1] popularized the GrabCut algorithm that describes the regions by highly descriptive distributions such as Gaussian mixture models (GMM) or histograms and uses iterated graph cuts to optimize a function of the general form (4.2). The authors focused on the problem of interactive foreground/background segmentation of a color image from a very simple initial user input, which amounts to a bounding box containing the region of interest (Fig. 4.1). This is of high interest in various applications, for instance, image editing. It can also be useful for synthesizing a large amount of masks for training semantic segmentation algorithms [27], assuming that the training images come with weak annotations such as boxes [27] or scribbles [28]. GrabCut can be viewed as an iterative extension of the popular Boykov–Jolly model [5], which uses a single graph cut, assuming known region histograms.

Let $\mathbf{u} = (u_p)_{p \in \Omega} \in \{0, 1\}^{|\Omega|}$ denote a binary vector, with u_p being an indicator variable for pixel p: $u_p = 1$ if pixel p belongs to the foreground region, S^1, and $u_p = 0$ within the background, S^0. Notice that, unlike the multi-label case, we omit, without ambiguity, the label index in the binary variables. To model the likelihoods $\Pr(\mathbf{f}_p | l, \boldsymbol{\theta})$ in (4.2), $l = 0, 1$, GrabCut uses two GMMs, one for the foreground and the other for the background, each taking the form of a mixture with K components. Furthermore, to handle GMMs in a computationally efficient way, one can introduce an additional variable vector $\mathbf{k} = (k_p)_{p \in \Omega} \in \{1, \ldots, K\}^{|\Omega|}$, with $k_p \in \{1, \ldots, K\}$ associating pixel p with one of the K mixture components, either from the foreground or the background probability. In this case, the GMM log-likelihood unary potentials can be written as follows, for each p and for $l \in \{0, 1\}$:

$$- \ln \Pr(\mathbf{f}_p | l, \boldsymbol{\theta}, k_p) \overset{c}{=} - \ln \pi^{l, k_p} + \mathcal{G}^{vect}(\mathbf{f}_p, \mathbf{m}^{l, k_p}, \boldsymbol{\Sigma}^{l, k_p})$$

$$= -\ln \pi^{l,k_p} + \frac{1}{2}\ln(\det \mathbf{\Sigma}^{l,k_p})$$

$$+ \frac{1}{2}(\mathbf{f}_p - \mathbf{m}^{l,k_p})^t(\mathbf{\Sigma}^{l,k_p})^{-1}(\mathbf{f}_p - \mathbf{m}^{l,k_p}). \quad (4.22)$$

The first term involves the mixture weighting parameters $\pi^{l,k}$, and the second is the logarithm of a Gaussian model, up to an additive constant. Thus, $\boldsymbol{\theta} = \{\boldsymbol{\theta}_0, \boldsymbol{\theta}_1\}$ is the set of model parameters, and $\boldsymbol{\theta}_l$ denotes the parameters of region l, $l \in \{0, 1\}$:

$$\boldsymbol{\theta}_l = \{\pi^{l,k}, \mathbf{m}^{l,k}, \mathbf{\Sigma}^{l,k}, k = 1 \ldots K\}. \quad (4.23)$$

In conjunction with the likelihood term, Rother et al. [1] used basic pairwise MRF potentials for boundary regularization and edge alignment:

$$\sum_{p,q \in N} w_{pq} \cdot [u_p \neq u_q], \quad (4.24)$$

where, recalling that [.] denotes the Iverson bracket, taking value 1 if its argument is true and 0 otherwise. Weights w_{pq} penalize discontinuities between pixels p and q:

$$w_{pq} = \lambda \exp\left(-\beta \|\mathbf{f}_p - \mathbf{f}_q\|^2\right). \quad (4.25)$$

Combining the GMM data term in (4.22) with regularization (4.24), the GrabCut function reads:

$$\mathcal{J}(\mathbf{u}, \boldsymbol{\theta}, \mathbf{k}) = -\sum_p u_p \ln \Pr(\mathbf{f}_p | 1, \boldsymbol{\theta}, k_p) - \sum_p (1 - u_p) \ln \Pr(\mathbf{f}_p | 0, \boldsymbol{\theta}, k_p)$$

$$+ \sum_{p,q \in N} w_{pq} \cdot [u_p \neq u_q]. \quad (4.26)$$

The main steps of the iterative GrabCut algorithm are summarized in the following:

1. Initialize segmentation $\mathbf{u}$;
2. Compute the initial foreground and background GMMs from the initial regions;
3. Iterate until convergence:
   a. Assign each pixel p to a GMM component of the current region of the pixel:

$$k_p = \underset{k_p}{\arg\max} \Pr(\mathbf{f}_p | u_p, \boldsymbol{\theta}, k_p). \quad (4.27)$$

b. Learn the GMM parameters from image data within the current regions:

$$\boldsymbol{\theta} = \arg\min_{\boldsymbol{\theta}} \mathcal{J}(\mathbf{u}, \boldsymbol{\theta}, \mathbf{k}). \tag{4.28}$$

c. Compute segmentation with a graph cut [5]:

$$\mathbf{u} = \arg\min_{\mathbf{u}} \mathcal{J}(\mathbf{u}, \boldsymbol{\theta}, \mathbf{k}). \tag{4.29}$$

Each of these steps decreases the function, which guarantees convergence. In the previous iterations, the first step (4.27) is simple: For each pixel p, it enumerates the k_p values and choose the best. The parameter learning step (4.28) can be implemented as follows. For each GMM component (l, k), we identify the region corresponding to the component:

$$S^{l,k} = \{p \mid k_p = k \text{ and } u_p = l\}. \tag{4.30}$$

Parameters $\mathbf{m}^{l,k}$ and $\boldsymbol{\Sigma}^{l,k}$ are estimated as in the case of a single Gaussian, using, respectively, the sample mean and covariance of color features within region $S^{l,k}$. The mixture weights are updated as follows:

$$\pi^{l,k} = \frac{|S^{l,k}|}{\sum_j |S^{l,j}|}. \tag{4.31}$$

The last step (4.29) optimizes $\mathcal{J}$ with respect to the segmentation. When the parameters are fixed, $\mathcal{J}$ takes the form of a sum of unary and submodular potentials. Therefore, a graph cut can be used.

Instead of GMMs, one can use normalized histograms as statistical descriptions of the image data within the regions. Let $\mathbf{z} : \mathbb{R}^N \to \mathcal{Z} \subset \mathbb{N}^N$ denotes a function that maps image features $\mathbf{f}_p$ to a finite set of bins, e.g., color bins. Then, the histogram of a given segmentation region S^l is computed as follows:

$$P_h(\mathbf{b}, l) = \frac{\sum_{p \in S^l} [\mathbf{z}(\mathbf{f}_p) = \mathbf{b}]}{|S^l|} \quad \forall \mathbf{b} \in \mathcal{Z}. \tag{4.32}$$

Essentially, $P_h(\mathbf{b}, l)$ counts the number of pixels of S^l belonging to bin $\mathbf{b}$, and normalizes such a bin count using the cardinality of region S^l. In the case of normalized histograms, likelihood $\Pr(\mathbf{f}_p | l, \boldsymbol{\theta})$ is estimated as $P_h(\mathbf{z}(\mathbf{f}_p), l)$ and parameter set $\boldsymbol{\theta}_l$ is given by $\{P_h(\mathbf{b}, l), \mathbf{b} \in \mathcal{Z}\}$.

References

[1] C. Rother, V. Kolmogorov, A. Blake, GrabCut: Interactive foreground extraction using iterated graph cuts, ACM Transactions on Graphics 23 (3) (2004) 309–314.

[2] T. Chan, L. Vese, Active contours without edges, IEEE Transactions on Image Processing 10 (2) (2001) 266–277.

[3] M.J. Kearns, Y. Mansour, A.Y. Ng, An information-theoretic analysis of hard and soft assignment methods for clustering, in: Proceedings of the Conference on Uncertainty in Artificial Intelligence (UAI), 1997, pp. 282–293.

[4] M.Á. Carreira-Perpiñán, Clustering methods based on kernel density estimators: Mean-shift algorithms, in: Handbook of Cluster Analysis, CRC/Chapman and Hall, 2016, pp. 383–418.

[5] Y. Boykov, M.P. Jolly, Interactive graph cuts for optimal boundary and region segmentation of objects in n-d images, in: IEEE International Conference on Computer Vision (ICCV), 2001, pp. 105–112.

[6] Y. Boykov, G. Funka Lea, Graph cuts and efficient n-d image segmentation, International Journal of Computer Vision 70 (2) (2006) 109–131.

[7] Y. Boykov, V. Kolmogorov, An experimental comparison of min-cut/max-flow algorithms for energy minimization in vision, IEEE Transactions on Pattern Analysis and Machine Intelligence 26 (9) (2004) 1124–1137.

[8] C.C. Aggarwal, C.K. Reddy (Eds.), Data Clustering: Algorithms and Applications, Chapman & Hall/CRC, 2014.

[9] S.C. Zhu, A.L. Yuille, Region competition: unifying snakes, region growing, and Bayes/MDL for multiband image segmentation, IEEE Transactions on Pattern Analysis and Machine Intelligence 18 (9) (1996) 884–900.

[10] A. Mitiche, I. Ben Ayed, Variational and Level Set Methods in Image Segmentation, 1st ed., Springer, 2011, 192 pp.

[11] R. Duda, P. Hart, D. Stork, Pattern Classification, John Wiley & Sons, 2001.

[12] C.M. Bishop, Pattern Recognition and Machine Learning, Springer, 2006.

[13] M. Tang, D. Marin, I. Ben Ayed, D. Marin, Y. Boykov, Kernel cuts: Kernel & spectral clustering meet regularization, International Journal of Computer Vision 127 (5) (2019) 477–511.

[14] M. Rousson, R. Deriche, A variational framework for active and adaptive segmentation of vector valued images, in: Workshop on Motion and Video Computing, 2002.

[15] L.A. Vese, T.F. Chan, A multiphase level set framework for image segmentation using the Mumford and shah model, International Journal of Computer Vision 50 (3) (2002) 271–293.

[16] M. di Bisceglie, C. Galdi, CFAR detection of extended objects in high-resolution SAR images, IEEE Transactions on Geoscience and Remote Sensing 43 (4) (2005) 833–843.

[17] I. Ben Ayed, A. Mitiche, Z. Belhadj, Multiregion level set partitioning of synthetic aperture radar images, IEEE Transactions on Pattern Analysis and Machine Intelligence 27 (5) (2005) 793–800.

[18] M. Mignotte, C. Collet, P. Perez, P. Bouthemy, Sonar image segmentation using an unsupervised hierarchical MRF model, IEEE Transactions on Image Processing 9 (7) (2000) 1216–1231.

[19] R. Archibald, J. Hu, A. Gelb, G.E. Farin, Improving the accuracy of volumetric segmentation using pre-processing boundary detection and image reconstruction, IEEE Transactions on Image Processing 13 (2004) 459–466.

[20] I. Ben Ayed, N. Hennane, A. Mitiche, Unsupervised variational image segmentation/classification using a Weibull observation model, IEEE Transactions on Image Processing 15 (11) (2006) 3431–3439.

[21] I.M. Ziko, E. Granger, I. Ben Ayed, Scalable Laplacian k-modes, in: Neural Information Processing Systems (NeurIPS), 2018, pp. 10062–10072.

[22] M.B. Salah, I. Ben Ayed, J. Yuan, H. Zhang, Convex-relaxed kernel mapping for image segmentation, IEEE Transactions on Image Processing 23 (3) (2014) 1143–1153.

[23] M. Ben Salah, A. Mitiche, I. Ben Ayed, Multiregion image segmentation by parametric kernel graph cuts, IEEE Transactions on Image Processing 20 (2) (2011) 545–557.

[24] B. Hwang, J. Ren, S. McCormack, C. Berry, I. Ben Ayed, H.C. Graber, et al., A practical algorithm for the retrieval of floe size distribution of arctic sea ice from high-resolution satellite synthetic aperture radar imagery, Elementa: Science of the Anthropocene 38 (5) (2017) 1–23.

[25] D. Comaniciu, P. Meer, Mean shift: A robust approach toward feature space analysis, IEEE Transactions on Pattern Analysis and Machine Intelligence 24 (5) (2002) 603–619.

[26] Y. Cheng, Mean shift, mode seeking, and clustering, IEEE Transactions on Pattern Analysis and Machine Intelligence 17 (8) (1995) 790–799.

[27] A. Khoreva, R. Benenson, J.H. Hosang, M. Hein, B. Schiele, Simple does it: Weakly supervised instance and semantic segmentation, in: IEEE Conference on Computer Vision and Pattern Recognition (CVPR), 2017, pp. 1665–1674.

[28] D. Lin, J. Dai, J. Jia, K. He, J. Sun, Scribblesup: Scribble-supervised convolutional networks for semantic segmentation, in: IEEE Conference on Computer Vision and Pattern Recognition (CVPR), 2016, pp. 3159–3167.

CHAPTER 5

Regularized mutual information

5.1 Model fitting as entropy minimization

In the previous chapter, we examined image segmentation models that integrate regularization and probabilistic model-fitting terms. Such very popular algorithms are motivated by maximum likelihood (ML) estimation. For segmentation into L regions, they minimize regularized model-fitting objectives that follow the general form:

$$\mathcal{J}(\mathbf{S}, \boldsymbol{\theta}) = -\sum_{l=0}^{L-1}\sum_{p \in S^l} \ln \Pr(\mathbf{f}_p | l, \boldsymbol{\theta}) + \lambda \mathcal{R}(\mathbf{S}), \qquad (5.1)$$

where $\mathbf{S} = \{S^l, l = 0, \ldots, L-1\}$ denotes a variable partition of spatial image domain Ω, and $\Pr(\mathbf{f}_p | l, \boldsymbol{\theta})$ a parametric probability model of observed features (e.g., intensity or colors) within segmentation region S^l, $l = 0, \ldots, L-1$. Variable $\boldsymbol{\theta}$ denotes a set of parameters characterizing these probabilities, and $\mathbf{f}_p \in \mathbb{R}^N$ is a vector containing the observed features at pixel p. The first term in (5.1) penalizes the deviation of image features within region S^l from the likelihood model $\Pr(. | l, \boldsymbol{\theta})$. The regularization term $\mathcal{R}$ encourages smooth, edge–aligned segmentation boundaries, as in the popular Potts model discussed in great detail earlier in the book.

For mixed objectives of the form (5.1), one typically pursue iterative two–step optimization [1–4]. One step fixes segmentation and finds the optimal set of parameters. Another step finds the optimal segmentation with the parameters fixed. As discussed in great detail in the previous chapter, the general model in (5.1) and its two–step optimization include several popular segmentation algorithms, such as GrabCut [3] or Chan–Vese [2], and have straightforward connections to the standard K-means clustering objective and its probabilistic variants [5].

This chapter gives several information-theoretic perspectives of these popular segmentation algorithms and the general regularized model-fitting in (5.1). In fact, image segmentation, or more generally data clustering, can be viewed as maximizing the mutual information between data points, i.e., the features, and a latent labeling [6,7]. First, in this section, we will examine the link between the general model fitting term in (5.1) and en-

High-Order Models in Semantic Image Segmentation
https://doi.org/10.1016/B978-0-12-805320-1.00010-5

tropy minimization via a generative view of the mutual information. This link shows that the standard model in (5.1) can be viewed as a regularized entropy minimization. Then, in the next section, we will examine a different, discriminative view of the mutual information that reveals interesting connections to discriminative clustering models [6]. These mutual information insights explain some artifacts that are observed in practice for popular ML-based segmentation algorithms, for instance, biases towards balanced solutions [8], and prescribe principled methodologies for dealing with such artifacts. We will support the discussion with several experiments.

In general, the mutual information evaluates the amount of dependence between two random variables. Let X be a random variable describing segmentation (or clustering). X takes its possible values in a finite set of discrete labels $\{0, \dots, L-1\}$. For instance, in the context of semantic segmentation of natural images, each label describes a semantic category, e.g., car, sky, bicycle, etc. Let $\mathbf{f}$ denotes a random variable describing feature vectors (data points), e.g., color or textures in the context of image segmentation. The mutual information between X and $\mathbf{f}$ can be written as follows [7]:

$$\mathcal{I}(X, \mathbf{f}) = \mathcal{H}(\mathbf{f}) - \mathcal{H}(\mathbf{f}|X), \tag{5.2}$$

where $\mathcal{H}(\mathbf{f})$ is the entropy of features and $\mathcal{H}(\mathbf{f}|X)$ the conditional entropy of features, which assumes the labels are given. The goal is to find a labeling that maximizes mutual information (5.2). $\mathcal{H}(\mathbf{f})$ is a constant independent of labeling and, therefore, can be ignored. The labeling is a discrete random variable whose marginal distribution can be estimated empirically by the proportion of points within each cluster or segment [7]:

$$\Pr(X = l) = \frac{|S^l|}{|\Omega|}. \tag{5.3}$$

Now, we can write the conditional entropy in (5.2) as follows:

$$\mathcal{H}(\mathbf{f}|X) = \sum_l \Pr(X = l)\mathcal{H}(\mathbf{f}|X = l) = \frac{1}{|\Omega|}\sum_l |S^l|\mathcal{H}(\mathbf{f}|X = l), \tag{5.4}$$

with each $\mathcal{H}(\mathbf{f}|X = l)$ given by:

$$\mathcal{H}(\mathbf{f}|X = l) = -\int_{\mathbf{f}} \Pr(\mathbf{f}/X = l)\ln\Pr(\mathbf{f}/X = l)d\mathbf{f}. \tag{5.5}$$

Hereafter, we will omit X and use $\mathcal{H}(\mathbf{f}|l)$ instead of $\mathcal{H}(\mathbf{f}|X=l)$, to simplify the notation. To estimate this conditional entropy, let us first consider the following well-known Monte Carlo estimation [5,9].

Proposition 1 (Monte Carlo estimation). *For any discrete subset of points $S \subset \Omega$ and any function g, we have:*

$$\int_{\mathbf{f}} g(\mathbf{f}) \Pr(\mathbf{f}|S) d\mathbf{f} \approx \frac{1}{|S|} \sum_{p \in S} g(\mathbf{f}_p), \qquad (5.6)$$

where $\Pr(\mathbf{f}|S)$ denotes the density of $\{\mathbf{f}_p, \ p \in S\}$.

Therefore, applying Monte Carlo to $\Pr(\mathbf{f}|S^l) \approx \Pr(\mathbf{f}|l, \boldsymbol{\theta})$ and $g(\mathbf{f}) = -\ln \Pr(\mathbf{f}|l, \boldsymbol{\theta})$ yields the following approximation of the conditional entropy in Eq. (5.5) for sufficiently descriptive models, e.g., histograms or Gaussian Mixture Models (GMMs):

$$\mathcal{H}(\mathbf{f}|l, \boldsymbol{\theta}) \approx -\frac{1}{|S^l|} \sum_{p \in S^l} \ln \Pr(\mathbf{f}_p|l, \boldsymbol{\theta}). \qquad (5.7)$$

Note that this approximation is exact for the case of using normalized histograms as regional probability models $\Pr(\mathbf{f}_p|l, \boldsymbol{\theta})$. Recall from Chapter 4 that the normalized histogram of segmentation region S^l is given by:

$$P_h(\mathbf{b}, l) = \frac{\sum_{p \in S^l} [\mathbf{z}(\mathbf{f}_p) = \mathbf{b}]}{|S^l|} \ \forall \, \mathbf{b} \in \mathcal{Z},$$

where $\mathbf{z}$ denotes a function mapping image features $\mathbf{f}_p$ to a finite set of bins $\mathcal{Z}$, $|S^l|$ is the cardinality of region S^l and [.] is the Iverson bracket, taking a value of 1 if its argument is true and 0 otherwise. Essentially, $P_h(\mathbf{b}, l)$ evaluates the proportion of pixels of S^l belonging to bin $\mathbf{b}$. In the case of normalized histograms, likelihood $\Pr(\mathbf{f}_p|l, \boldsymbol{\theta})$ is estimated as $P_h(\mathbf{z}(\mathbf{f}_p), l)$, and parameter set $\boldsymbol{\theta}$ is given by the set of values $\{P_h(\mathbf{b}, l)\}$, for $l \in \{0, \ldots, L-1\}$ and $\mathbf{b} \in \mathcal{Z}$. In this case, and to see that approximation (5.7) is exact, it suffices to replace the regional summation over pixels in (5.7) by a summation over histogram bins:

$$-\sum_{p \in S^l} \ln P_h(\mathbf{z}(\mathbf{f}_p), l) = -\sum_{\mathbf{b} \in \mathcal{Z}} |S^l| P_h(\mathbf{b}, l) \ln P_h(\mathbf{b}, l) = |S^l| \mathcal{H}(\mathbf{f}|l). \qquad (5.8)$$

Combining (5.7), (5.4), and (5.2), it is easy to see that the standard model fitting term in (5.1) is an approximation (up to additive and multiplicative constants) to the negative mutual information, which can be expressed in terms of the entropies of features within the regions:

$$- |\Omega| \mathcal{I}(X, \mathbf{f}) \stackrel{c}{=} \sum_{l} |S^l| \mathcal{H}(\mathbf{f}|l, \boldsymbol{\theta}) \approx - \sum_{l} \sum_{p \in S^l} \ln \Pr(\mathbf{f}_p|l, \boldsymbol{\theta}). \qquad (5.9)$$

The entropy criterion in (5.9) is well known in the literature, both in the context of data clustering [5,10] as well as in image segmentation [11,12]. In fact, the authors of [5] examined the model-fitting term in (5.1) as a general clustering criterion, which they referred to as *probabilistic* K-means. They showed the link between probabilistic K-means and the entropy criterion in (5.9). In general, the entropy evaluates the amount of uncertainty (or randomness) associated with a distribution. Therefore, each conditional entropy in (5.9) evaluates the amount of heterogeneity of features within segmentation region S^l. Homogeneous regions correspond to low entropies and high mutual-information values. Intuitively, this means that minimizing the entropy prefers regions having peaked (or tight) distributions.

5.2 Limitations of entropy and highly descriptive models

In the following, let us examine the limitations of the entropy segmentation (or clustering) criterion in the context of highly descriptive models such as histograms or GMMs. Fig. 5.1 depicts the key problem for clustering color features with histograms. In this example, the second and third columns show two different segmentations of the image in the first column. The image and segmentations are depicted in the upper row. The bottom row shows the image histogram, with the corresponding segmentation regions coded in colors. It is clear that the entropy does not differentiate between the two solutions: Any bin permutation does not change the entropy. In fact, when a continuous color space is mapped to a set of finite bins, the entropy of histograms does not account for information on proximity between colors. The probabilities of nearby colors might be completely different.

The problem is different in the case of using GMMs with the entropy. In general, continuous density estimation accounts for the similarity between colors, with nearby color-space features typically having close probabilities. The difficulty with GMMs is related to the optimization problem. As discussed in Chapter 4 in the case of K-means, one can express the functional solely in term of segmentation variables S^l by replacing the set of parameters, i.e., variable $\boldsymbol{\theta}$, with the corresponding optimal values. This yields

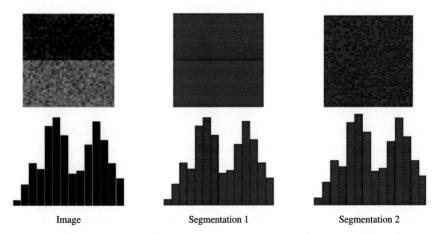

Image   Segmentation 1   Segmentation 2

Figure 5.1 Entropy of histograms. The entropy of histograms does not account for information about proximity between colors. It also does not differentiate between the two solutions in the second and third columns. (*Figures from [13] © IEEE.*)

high-order fractional terms. Although GMMs and GrabCut are very successful in the context of color segmentation, the use of complex, highly descriptive models introduces a large number of variable parameters, typically leading to complex optimization problems. In this case, alternating optimization schemes, with one step finding the optimal regions and the other minimizing with respect to the parameters, are likely to get trapped at weak local minima. Fig. 5.2 illustrates this with a simple example of clustering two–dimensional data. Fig. 5.2 (c) shows a good solution when using the ground truth as initialization and choosing a specific number of GMM components. However, changing slightly the initialization to the one in (a) over-fitted the data, as depicted in (d), yielding a higher value of the functional at convergence. This corresponds to a local minimum. Notice that the choice of the number of GMM components is important, and the use of a single component for each region, as depicted in (b), cannot find a good clustering. This sensitivity to local minima might be further accrued in the case of higher-dimensional feature spaces [9], which might explain why this class of clustering techniques is not common beyond the context of segmenting color feature spaces ($\mathbf{f}_p \in \mathbb{R}^3$). In fact, within the learning community, kernel methods [10] are widely used for partitioning high-dimensional features. Such kernel-based techniques also involve high-order terms, which will be discussed in the next chapter.

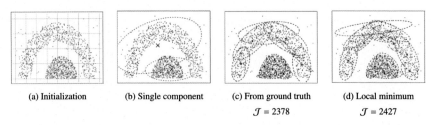

(a) Initialization   (b) Single component   (c) From ground truth   (d) Local minimum

$\mathcal{J} = 2378$ $\mathcal{J} = 2427$

Figure 5.2 Illustration of the difficulties with GMMs and entropy. (a) Initialization for partitioning two-dimensional data into two clusters, via model fitting (5.1) with GMM-based likelihoods and $\lambda = 0$ (no regularization); (b) The use of a single mixture component for each region cannot find a good clustering; (c) A good solution when starting from the ground truth as initialization and choosing a specific number of GMM components; (d) Local minimum when using the initialization in (a), changing slightly the ground-truth one used in (c). (*Figures from [13] © IEEE.*)

5.3 A discriminative view of the mutual information

5.3.1 Bias towards balanced segmentation

In this section, we will examine another information–theoretic perspective, which reveals that the standard likelihood term in (5.1) has an intrinsic bias towards balanced solutions, i.e., clusters (or segments) of equal size. Such a bias is well known in machine learning [5]. In the context of image segmentation, it might yield significant artifacts in practice [8]; see the example in Fig. 5.3. Furthermore, the analysis here shows a connection between discriminative clustering objectives [6] and the generative model fitting in (5.1).

We saw earlier that the likelihood term in (5.1) is an approximation of the mutual information between the features and the sought labeling. The approximation follows from expression (5.2) of the mutual information. Now, notice that we can rewrite the latter in another way [6]:

$$\mathcal{I}(X, \mathbf{f}) = \mathcal{H}(X) - \mathcal{H}(X|\mathbf{f}). \tag{5.10}$$

Combining (5.10) and (5.9), we get the following:

$$-\sum_{l}\sum_{p \in S^l} \ln \Pr(\mathbf{f}_p | l, \boldsymbol{\theta}) \approx |\Omega| \left(\mathcal{H}(X|\mathbf{f}) - \mathcal{H}(X) \right). \tag{5.11}$$

(a) Target proportion   (b) Without bias   (c) Standard likelihoods

Figure 5.3 Illustration of the label-proportion biases in model fitting. (a) Segmentation with the known distribution of labels (evaluated from the ground truth separating the person from the background); (b) The solution obtained by adding $|\Omega|\mathcal{H}(X)$ to the likelihood term, removing completely the bias; (c) The solution obtained with the likelihood term and the GrabCut algorithm, with the initial foreground being the whole image domain. The obtained segmentation corresponds to a very balanced partition, which did not separate the person from the background. (*Figures from [8] © IEEE.*)

Conditional entropy $\mathcal{H}(X|\mathbf{f})$ can be approximated with Monte Carlo estimation (5.6):

$$\mathcal{H}(X|\mathbf{f}) = \int_{\mathbf{f}} \Pr(\mathbf{f})\mathcal{H}(X|\mathbf{f})d\mathbf{f} \approx \frac{1}{|\Omega|}\sum_{p\in\Omega}\mathcal{H}(X|\mathbf{f}_p), \qquad (5.12)$$

where $\mathcal{H}(X|\mathbf{f}_p)$ is the entropy of posteriors for each data point $\mathbf{f}_p$:

$$\mathcal{H}(X|\mathbf{f}_p) = -\sum_{l}\Pr(l|\mathbf{f}_p,\boldsymbol{\theta})\ln\Pr(l|\mathbf{f}_p,\boldsymbol{\theta}). \qquad (5.13)$$

Minimizing this conditional entropy for a given data point $\mathbf{f}_p$ minimizes the uncertainty associated to assigning labels to the point. This point-wise entropy reaches its minimum when a single label k has the maximum posterior, i.e., $\Pr(k|\mathbf{f}_p,\boldsymbol{\theta}) = 1$, whereas each of the other labels verifies $\Pr(j|\mathbf{f}_p,\boldsymbol{\theta}) = 0$, $j \neq k$. It reaches its maximum when all the posteriors are equal, i.e., $\Pr(l|\mathbf{f}_p,\boldsymbol{\theta}) = 1/L \;\forall l$, which corresponds to the highest level of uncertainty.

The term $-|\Omega|\mathcal{H}(X)$ in approximation (5.11) is the one responsible for biasing the solutions towards balanced partitions. This term corresponds to minimizing, up to a positive multiplicative constant, the negative entropy of the marginal distribution of labels. The latter can estimated as follows:

$$-\mathcal{H}(X) = \sum_l \Pr(X = l) \ln \Pr(X = l) \approx \sum_l \frac{|S^l|}{|\Omega|} \ln \frac{|S^l|}{|\Omega|}. \tag{5.14}$$

Notice that we can write, up to a constant, this term as the following Kullback–Leibler (KL) divergence between the marginal probabilities of labels and the uniform distribution $\mathcal{U} = \{\frac{1}{L}, ..., \frac{1}{L}\}$:

$$\text{KL}\left(\Pr(X = l) \| \mathcal{U}\right) = \sum_l \frac{|S^l|}{|\Omega|} \ln \frac{|S^l|/|\Omega|}{1/L}. \tag{5.15}$$

Minimizing this KL divergence encourages solutions with equal-size segments. The minimum of this term is reached for solutions verifying

$$|S^l| = \frac{\Omega}{L} \ \forall l.$$

Therefore, approximation (5.11) shows that the standard likelihood term has an intrinsic bias towards balanced solutions because its optimization yields an approximate solution for minimizing:

$$|\Omega|\left(\mathcal{H}(X|\mathbf{f}) + \text{KL}\left(\Pr(X = l) \| \mathcal{U}\right)\right). \tag{5.16}$$

Fig. 5.3 (c) depicts how this balanced-solution bias might be strong in practice, yielding significant artifacts in image segmentation [8].

In fact, the combination of the conditional entropy of posteriors in (5.13) and balancing term (5.14) is common in the context of discriminative clustering [6,14]. In this case, one has to choose a parametric model for posteriors $\Pr(l|\mathbf{f}_p, \boldsymbol{\theta})$, e.g., unsupervised multilogit regression [6]. In the context of unsupervised image segmentation, it might be more convenient to specify the likelihoods [15], particularly when we have an idea about the distribution of image data within the segmentation regions, as is the case in many applications. Therefore, to control the effect of the bias, we can add an extra term $\gamma|\Omega|\mathcal{H}(X)$ to the likelihoods [8], with γ a positive constant. Setting $\gamma = 1$ compensates exactly for the balanced-solution bias term in (5.11), thereby removing it completely. Values of γ in $]0, 1[$ reduce the effect of the bias without removing it. Fig. 5.3 (c) shows the solution obtained with the likelihood term and the GrabCut algorithm. In this example, the initial foreground is the whole image domain, and the obtained solution corresponds to a very balanced partition but did not separate the person from the background. Adding term $|\Omega|\mathcal{H}(X)$, which completely removes the bias, yielded the solution in Fig. 5.3 (b).

Notice that $\mathcal{H}(X)$ is a high-order term, and subsequent chapters will address in-depth details on several efficient high-order optimization techniques. For now, it is worth noting, however, that $\mathcal{H}(X)$ is a concave function of cardinality, a special structure that makes its exact optimization easy for binary discrete problems. In fact, it is well known that, for binary problems, concave cardinality functions are submodular [16]. Such functions were used and optimized exactly in several applications in computer vision [8,11,17], including segmentation.

5.3.2 Bias to a target label distribution

Expression (5.16) reveals that we can replace uniform distribution $\mathcal{U}$ with any target (fixed) label distribution [6,8] $\mathcal{W} = \{w^1, w^2, ..., w^L\}$, which gives the following KL divergence:

$$\mathrm{KL}\left(\mathrm{Pr}(X = l) \,||\, \mathcal{W}\right) = \sum_l \frac{|S^l|}{|\Omega|} \ln \frac{|S^l|/|\Omega|}{w^l}. \tag{5.17}$$

In fact, imposing such a prior on the proportion of feature points within each of the clusters is standard in the context of discriminative clustering [6,14]. Also, in segmentation, such a prior can be very helpful [8], and is easy to add from the optimization perspective. Notice that, using (5.11), a weighted version of the likelihood term can be approximated as follows:

$$-\sum_l \sum_{p \in S^l} \ln\left(w^l \, \mathrm{Pr}(\mathbf{f}_p | l, \boldsymbol{\theta})\right) = -\sum_l |S^l| \ln w^l - \sum_l \sum_{p \in S^l} \mathrm{Pr}(\mathbf{f}_p | l, \boldsymbol{\theta})$$

$$\approx -\sum_l |S^l| \ln w^l + |\Omega| \left(\mathcal{H}(X|\mathbf{f}) - \mathcal{H}(X)\right)$$

$$= |\Omega| \left(\mathcal{H}(X|\mathbf{f}) + \sum_l \frac{|S^l|}{|\Omega|} \ln \frac{|S^l|/|\Omega|}{w^l}\right)$$

$$= |\Omega| \left(\mathcal{H}(X|\mathbf{f}) + \mathrm{KL}\left(\mathrm{Pr}(X = l) \,||\, \mathcal{W}\right)\right). \tag{5.18}$$

Therefore, when we have a known prior $\mathcal{W}$, the KL term appearing in the last line of (5.18) means that the weighted likelihood functional has an intrinsic bias to a known label-proportion distribution $\mathcal{W}$. It is worth noting that using weighted likelihoods does not result in any additional difficulty or computations from optimization perspective. These weights add unary potentials whose optimization is trivial. However, as illustrated in Fig. 5.3 (a), if the target proportion of labels is known, optimizing the weighted likelihoods can be viewed as an indirect optimization of the KL divergence in (5.17), which improves the performance.

References

[1] S.C. Zhu, A.L. Yuille, Region competition: Unifying snakes, region growing, and Bayes/MDL for multiband image segmentation, IEEE Transactions on Pattern Analysis and Machine Intelligence 18 (9) (1996) 884–900.

[2] T. Chan, L. Vese, Active contours without edges, IEEE Transactions on Image Processing 10 (2) (2001) 266–277.

[3] C. Rother, V. Kolmogorov, A. Blake, GrabCut: Interactive foreground extraction using iterated graph cuts, ACM Transactions on Graphics 23 (3) (2004) 309–314.

[4] A. Mitiche, I. Ben Ayed, Variational and Level Set Methods in Image Segmentation, 1st ed., Springer, 2011, 192 pp.

[5] M.J. Kearns, Y. Mansour, A.Y. Ng, An information–theoretic analysis of hard and soft assignment methods for clustering, in: Proceedings of the Conference on Uncertainty in Artificial Intelligence (UAI), 1997, pp. 282–293.

[6] R. Gomes, A. Krause, P. Perona, Discriminative clustering by regularized information maximization, in: Advances in Neural Information Processing Systems (NIPS), 2010, pp. 775–783.

[7] J. Kim, J.W. Fisher III, A.J. Yezzi, M. Çetin, A.S. Willsky, A nonparametric statistical method for image segmentation using information theory and curve evolution, IEEE Transactions on Image Processing 14 (10) (2005) 1486–1502.

[8] Y. Boykov, H.N. Isack, C. Olsson, I. Ben Ayed, Volumetric bias in segmentation and reconstruction: Secrets and solutions, in: IEEE International Conference on Computer Vision (ICCV), 2015, pp. 1769–1777.

[9] M. Tang, D. Marin, I. Ben Ayed, D. Marin, Y. Boykov, Kernel cuts: Kernel & spectral clustering meet regularization, International Journal of Computer Vision 127 (5) (2019) 477–511.

[10] D. Marin, M. Tang, I. Ben Ayed, Y. Boykov, Kernel clustering: Density biases and solutions, IEEE Transactions on Pattern Analysis and Machine Intelligence 41 (1) (2018) 136–147.

[11] M. Tang, L. Gorelick, O. Veksler, Y. Boykov, GrabCut in one cut, in: International Conference on Computer Vision (ICCV), 2013, pp. 1769–1776.

[12] M. Tang, I. Ben Ayed, Y. Boykov, Pseudo-bound optimization for binary energies, in: European Conference on Computer Vision (ECCV), Part V, 2014, pp. 691–707.

[13] M. Tang, I. Ben Ayed, D. Marin, Y. Boykov, Secrets of GrabCut and kernel k-means, in: IEEE International Conference on Computer Vision (ICCV), 2015, pp. 1555–1563.

[14] K.G. Dizaji, A. Herandi, C. Deng, W. Cai, H. Huang, Deep clustering via joint convolutional autoencoder embedding and relative entropy minimization, in: IEEE International Conference on Computer Vision (ICCV), 2017, pp. 5747–5756.

[15] D. Cremers, M. Rousson, R. Deriche, A review of statistical approaches to level set segmentation: Integrating color, texture, motion and shape, International Journal of Computer Vision 72 (2) (2007) 195–215.

[16] L. Lovasz, Submodular functions and convexity, in: Mathematical Programming: The State of the Art, 1983, pp. 235–257.

[17] P. Kohli, L. Ladicky, P.H.S. Torr, Robust higher order potentials for enforcing label consistency, International Journal of Computer Vision 82 (3) (2009) 302–324.

CHAPTER 6

Examples of high-order functionals

6.1 Introduction

Image segmentation methods are commonly built on minimizing functionals that take the general form:

$$\mathcal{G}(\mathbf{u}) + \mathcal{R}(\mathbf{u}), \tag{6.1}$$

where $\mathbf{u} = (u_p)_{p \in \Omega} \in \{0, 1\}^{|\Omega|}$ is a binary vector containing indicator variables of a target region $S \in \Omega$ (the foreground): $u_p = 1$ if pixel p belongs to S, and $u_p = 0$ otherwise. $\Omega \subset \mathbb{R}^2$ denotes the image domain. $\mathcal{R}(\mathbf{u})$ is an MRF/CRF boundary regularizer, and $\mathcal{G}(\mathbf{u})$ typically imposes desirable regional properties on the sought segmentation regions. To simplify the presentation in this chapter, we consider only the case of segmentation into two regions, a foreground S and a background $\Omega \setminus S$. The simplest form of a regional term $\mathcal{G}(\mathbf{u})$ is unary (or linear), which corresponds to summations over individual-pixel penalties [1,2]:

$$\mathcal{G}(\mathbf{u}) = \mathcal{L}(\mathbf{u}) = \sum_{p \in \Omega} u_p v_p = \mathbf{u}^t \mathbf{v}, \tag{6.2}$$

where $\mathbf{v} = (v_p)_{p \in \Omega} \in \mathbb{R}^{|\Omega|}$ is a vector containing a fixed potential v_p for each p.

Minimization of a unary term of the form in Eq. (6.2) alone, i.e. without regularizer $\mathcal{R}(\mathbf{u})$, is trivial. Indeed, one could treat each binary variable u_p independently of the other binary variables u_q, $q \neq p$: Given the fact that $u_p \in \{0, 1\}$, each term $u_p v_p$ in the summation reaches its smallest possible value for the following solution: $u_p = 1$ if v_p is negative and $u_p = 0$ otherwise.

The popular Boykov–Jolly model [1,2], which we described in great detail in Chapter 1, uses a linear regional term of the form in (6.2), with unary potentials v_p given by:

$$v_p = -\ln \frac{\Pr(\mathbf{f}_p | 1, \boldsymbol{\theta})}{\Pr(\mathbf{f}_p | 0, \boldsymbol{\theta})},$$

where $\Pr(\mathbf{f}_p|l, \boldsymbol{\theta})$ are fixed probability models for the foreground ($l = 1$) and background ($l = 0$) regions. $\mathbf{f}_p \in \mathbb{R}^N$, $p \in \Omega$, denotes a vector containing image features, e.g., color, intensity or textures, and $\boldsymbol{\theta}$ is a set of parameters. Typically, regional unary terms are used jointly with pairwise boundary regularizers such as the Potts model [3–5], which we also discussed in great detail in Chapter 1. The ensuing overall objective functions are submodular and, therefore, amenable to powerful combinatorial optimizers such as graph cuts [6], with optimality bounds or guarantees. Those sub-modular optimization aspects, quite instrumental and widely used in computer vision, were also discussed in Chapter 2.

Unfortunately, the set of properties that unary terms can impose on the solutions are quite limited. Therefore, high-order functions were intensively investigated in the MRF literature [7–15]. Furthermore, very recently, high-order functions in MRFs motivated several interesting loss functions and priors in the context of deep learning models [16–19], either during training [16,18] or during inference [17,20]. In this setting, the probability outputs of a deep network could be viewed as soft, parameterized versions of the discrete (binary) indicator variables of a segmentation region in the context of classical MRFs. Unlike linear terms (6.2), a high-order objective is a nonlinear function of one or many linear and/or pairwise terms. A quite general form of a high-order function could be written as follows:

$$\mathcal{G}(\mathbf{u}) = g(\mathbf{u}^t \mathbf{v}_1, \ldots, \mathbf{u}^t \mathbf{v}_M; \mathbf{u}^t W_1 \mathbf{u}^t, \ldots, \mathbf{u}^t W_N \mathbf{u}^t), \qquad (6.3)$$

where $g : \mathbb{R}^{M+N} \to \mathbb{R}$ is a nonlinear multivariate function, $\mathbf{v}_m \in \mathbb{R}^{|\Omega|}$, $m = 1, \ldots M$, are fixed vectors of unary potentials, and W_n, $n = 1, \ldots N$, are $|\Omega|$-by-$|\Omega|$ matrices containing fixed pairwise potentials. For instance, $\mathcal{G}$ may take the form of a ratio of pairwise to unary terms, as is the case of the graph clustering and shape prior objectives discussed in the following.

Unlike the summation of unary terms in Eq. (6.2), the solution for each binary variable u_p in (6.3) depends on all the other variables u_q, $q \neq p$, when g is a nonlinear function, e.g., a ratio function. Therefore, optimizing high-order functions of the form (6.3) is, in general, difficult. It often requires approximate and iterative solutions. This chapter will not discuss the optimization aspects of high-order functions. A few quite general optimization options will be discussed in great detail in Chapters 7 and 8. Here, we will give a few examples of high-order functions and discuss their practical use in image segmentation. Of course, this chapter is by no means a comprehensive survey of high-order functions in computer vision and machine learning. Instead, the goal is to show a few examples and how they embed useful priors on the solutions. This includes priors on the shapes of the segmentation regions [13,21], which can be very useful in medical-imaging applications [16,17,22]. This also includes priors the distributions of features within the target regions as well as graph-clustering objectives such the very popular normalized cut [23,24].

The treatment we will provide in the following will focus on binary variables and high-order objective that originated from the literature on discrete MRFs. These also have motivated soft, parameterized objectives, which were recently shown to embed very useful priors in the context of deep networks, during both the training [16,18] and inference phases [17]. However, we will not treat these deep-learning aspects in this chapter; we will dedicate two chapters at the end of the book (Chapters 9 and 10), where we discuss examples of high-order and pairwise MRF functions based on soft, parameterized segment-indicator variables in the form of deep-network outputs. In these cases, optimization is carried out with gradient descent, the workhorse for training deep networks.

Finally, it is worth pointing out that we already saw a few examples of high-order functions earlier in the book. Specifically, Chapter 4 focused on likelihood-based, model-fitting objectives for both segmentation and data clustering, which include the widely used K-means and its probabilistic generalizations. A breadth of segmentation algorithms, such as the very popular GrabCut [25] and Chan–Vese [26] models, could be viewed as spatially-regularized versions of probabilistic K-means clustering. In these models, optimization is carried out with respect to both the segmentation variables and the parameters of the probability distributions modeling the data within each segmentation region. The objective functions are high-order due to the nonlinear dependence between each segment S and its set of distribution parameters θ_S. In fact, for some forms of probability distributions, one can express the likelihoods solely in terms of the segmentation regions by replacing the parameters with the corresponding optimal, segmentation-dependent values. Even more generally, and as discussed in Chapter 5, optimizing the likelihood terms is equivalent to optimizing approximations of the segmentation-region entropy multiplied by region size. Chapter 5 also examined an information-theoretic perspective of popular likelihood segmentation algorithms, revealing biases towards balanced partitions and pointing to interesting high-order, region-size terms that could mitigate such biases.

6.2 Shape priors

In medical imaging, it is very common to have access to some prior knowledge about the size or shape of the target regions. For instance, approximate region size information might be available in the radiology text reports associated with the images. Such knowledge does not have to be precise, and may take the imprecise form of lower and upper bounds on segmentation-region sizes and shapes [13,16,17,22]. Therefore, imposing such priors on segmentation objectives, in the form of high–order functions, could mitigate the difficulty of image segmentation. In the context of discrete MRFs, several works showed how to integrate such high-order priors in standard regularization-based objective functions, e.g., [13]. This may take the following general constrained-optimization form:

$$\min_{\mathbf{u}} \quad \mathcal{L}(\mathbf{u}) + \mathcal{R}(\mathbf{u})$$

$$\text{s.t.} \quad g_m(\mathbf{u}) \leq 0, \ m = 1, \dots M, \tag{6.4}$$

where $\mathcal{L}(\mathbf{u})$ is unary term, $\mathcal{R}(\mathbf{u})$ is a pairwise regularization and inequality constraints $g_m(\mathbf{u}) \leq 0$, $m = 1, \dots M$ impose some prior knowledge about the sought solutions.

6.2.1 Region-size prior

The inequality constraints in (6.4) integrate priors that may stem from some domain-specific knowledge about the target regions. Assume, for instance, that we have some knowledge about the size of the target foreground region $S = \{p \in \Omega \mid u_p = 1\}$. Such knowledge does not have to be very precise and could take the form of lower and upper bounds on region size, a practical scenario that occur frequently in medical image segmentation problems [13, 16,17,22,27]. For example, imposing an upper bound a on the foreground region size could be done via the following constraint:

$$\sum_{p \in \Omega} u_p - a \leq 0.$$

In this case, the corresponding constraint in the general expression in Eq. (6.4) reads

$$g_1(\mathbf{u}) = \sum_{p \in \Omega} u_p - a.$$

Similarly, one can impose a lower bound b on region size by using

$$g_2(\mathbf{u}) = b - \sum_{p \in \Omega} u_p.$$

One way to address general constrained problem (6.4) is to follow a penalty-based approach that approximates the original constrained problem with an unconstrained one:

$$\min_{\mathbf{u}} \mathcal{L}(\mathbf{u}) + \mathcal{R}(\mathbf{u}) + \gamma \mathcal{P}(\mathbf{u}), \tag{6.5}$$

with γ is a positive hyper-parameter, and $\mathcal{P}$ is a positive, continuous and differentiable function verifying: $\mathcal{P}(\mathbf{u}) = 0$ if and only if $\mathbf{u}$ satisfies all the constraints. Penalty $\mathcal{P}$ increases when a constraint is not satisfied. Therefore, minimizing term $\mathcal{P}(\mathbf{u})$ in (6.5) encourages satisfaction of the inequality constraints of the original problem in (6.4). One possible choice for $\mathcal{P}$ would to use a quadratic penalty for imposing upper and lower bounds on region size [16,28]:

$$\mathcal{P}(\mathbf{u}) = [g_1(\mathbf{u})]_+^2 + [g_2(\mathbf{u})]_+^2, \tag{6.6}$$

where $[x]_+ = \max(0, x)$ denotes the rectifier function. Obviously, this choice of quadratic penalties for handling the constraints is not unique, and there are other options [29].

Fig. 6.1 depicts an example of liver segmentation in a Computed To-mography (CT) image. The left side of the figure shows segmentation obtained with the standard Boykov–Jolly model, i.e., by minimizing an objective containing a unary log-likelihood term based on intensity and a pairwise MRF boundary regularization. Due to the similarity in the intensity profiles of the target region (i.e., the liver structure) and it surrounding regions in the background, the Boykov–Jolly model yielded an over-segmentation. The left side of the figure depicts the solution obtained by adding a size constraint that corrected the result of the Boykov–Jolly model.

6.2.2 Shape moments

Beyond region size, one could further embed priors on the shapes of the target segmentation regions. Shape-moment descriptors provide a convenient tool to do so [13,17,19,30,31]. Shape moments of a segmentation

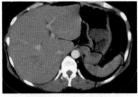
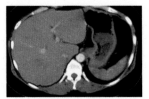

Without size prior   With size prior

Figure 6.1 Example of liver segmentation in a Computed Tomography (CT) image. Left: Result obtained with the standard Boykov–Jolly model; Right: Result obtained by adding a size constraint.

region $S = \{p \in \Omega \,|\, u_p = 1\}$ could be written as functions of segmentation variable $\mathbf{u}$ as follows:

$$m_{i,j}(\mathbf{u}) := \sum_{p \in \Omega} u_p x_p^i y_p^j, \tag{6.7}$$

where $i, j \in \mathbb{N}$ are the moment orders and x_p and y_p denote the spatial coordinates of pixel $p \in \Omega$, i.e. $p = (x_p, y_p)$. Notice that $m_{0,0}(\mathbf{u})$ corresponds exactly to region size. In fact, the shape moments of small orders i and j describe global information about a given shape, such as size, centroid and circularity. Larger shape-moment orders contain fine-grained information about the shape. It is worth noting that a few geometric shape moments could be sufficient to reconstruct complex shapes [32], via solving an inverse problem. Furthermore, such geometric shape moments could be made invariant with respect to geometric transformations (e.g., rotation, translation and scaling) by pure mathematical manipulations, which is convenient for segmentation [30]. This includes the well-known Hu's invariant moments [33]. For instance, to make shape moments translation-invariant, it suffices to shift pixel coordinates x_p and y_p by shape centroid, which yields the central moments:

$$\bar{m}_{i,j}(\mathbf{u}) = \sum_{p \in \Omega} u_p \left(x_p - \frac{m_{1,0}(\mathbf{u})}{m_{0,0}(\mathbf{u})} \right)^i \left(y_p - \frac{m_{0,1}(\mathbf{u})}{m_{0,0}(\mathbf{u})} \right)^j. \tag{6.8}$$

Shape-moment priors could be very useful, both in classical interactive-segmentation scenarios [30] and in recent deep-learning based solutions [17]. One could impose such priors by following the same constrained formulation we discussed in the previous subsection for imposing region-size constraints.

6.2.3 Shape compactness

In Chapter 1, we discussed in great detail classical MRF regularization, which takes the form of summations of pairwise potentials. We emphasized the Potts model as a widely used pairwise regularizer:

$$\mathcal{R}(\mathbf{u}) = \lambda \sum_{p,q \in \mathcal{N}} [u_p \neq u_q], \qquad (6.9)$$

where $\mathcal{N}$ denotes the set of pairs of neighboring pixels (e.g., a 4-, 8-, or 16-neighborhood system), λ is a strictly positive constant and [.] the Iverson bracket, which is equal to 1 if its argument is true and 0 otherwise. Minimization of (6.9) encourages nearby pixels to have the same label: We pay a penalty λ when two neighboring pixels have different labels. In the binary-segmentation case, it is straightforward to see that the Potts objective in (6.9) is proportional to the length of the segmentation boundary. As a result, minimizing this term acts as a smoothness prior, which encourages regular segmentation boundaries.

Clearly, the standard nonnegative boundary-length objective in (6.9) has a trivial minimum that corresponds to a zero-length boundary $\mathcal{R}(\mathbf{u}) = 0$ and an empty foreground segmentation region. In fact, the Potts prior is never used alone as the objective; it is often integrated with data-fidelity terms. However, the fact that (6.9) is small for low–cardinality segmentation regions means that models integrating Potts are biased towards small regions. This bias is well known as the *shrinking problem* [34].

One way to mitigate the shrinking problem is to use the ratio of boundary length to region size as a regularization [34,35]:

$$\mathcal{R}(\mathbf{u}) = \frac{\lambda \sum_{p,q \in \mathcal{N}} [u_p \neq u_q]}{\sum_{p \in \Omega} u_p}. \qquad (6.10)$$

This regularizer is the ratio of a pairwise term to a unary term and is, therefore, a high-order function. Optimizing this ratio function over binary variables is an NP-hard problem [34]. It is related to the well-known isoperimetric graph-partitioning problem, also referred to as the Cheeger problem [35]. Also, this regularizer is closely related to popular graph clustering objectives such as normalized cut, which we will discuss subsequently. Unlike the standard length regularizer in (6.9), the length-to-area ratio in (6.10) does not suffer from strong shrinking bias (i.e., bias towards small regions). However, as observed experimentally in several works

[34,35], isoperimetric partitioning has a strong bias towards large regions, which might be a serious limitation in image segmentation.

One way to avoid the bias of isoperimetric partitioning towards large regions is to minimize the following length-squared to area ratio function as a segmentation regularizer [21]:

$$\mathcal{R}(\mathbf{u}) = \frac{\lambda \left(\sum_{p,q \in \mathcal{N}} [u_p \neq u_q] \right)^2}{\sum_{p \in \Omega} u_p}. \qquad (6.11)$$

Notice that the ratio in (6.11) is dimensionless and, therefore, unbiased. In fact, minimizing (6.11) is equivalent to maximizing a shape circularity (or compactness) measure, which is widely used in the context of shape analysis [36]:

$$\frac{4\pi \sum_{p \in \Omega} u_p}{\left(\sum_{p,q \in \mathcal{N}} [u_p \neq u_q] \right)^2}. \qquad (6.12)$$

Measure (6.12) is in $(0, 1]$, and is equal to 1 if and only if the shape corresponding to binary variable $\mathbf{u}$ is a circle. The measure assesses shape compactness since it evaluates the deviation of a given shape from the most compact shape (i.e., a circle); the higher the value of (6.12), the closer the shape of the segmentation defined by $\mathbf{u}$ to a circle.

Shape compactness is a generic shape prior that could be very useful in image segmentation. This is particularly the case in medical-imaging applications, as illustrated in Fig. 6.2. In medical imaging, it is often the case that the target regions have compact shapes but weak (low-contrast) image edges. Of course, (6.11) could not be used alone as the objective since any circular region within the image domain would minimize globally this term, independently of image content. Therefore, this term should be used along with data-fidelity terms. In this case, (6.11) encourages shape compactness but does not yield necessarily perfectly circular shapes; see, for instance, the first row in Fig. 6.2.

One could integrate (6.11) with a data-fitting term that takes the form of unary potentials, as in generic expression (6.2). The work in [21] showed the positive effect of correcting the probability outputs of a deep network with the shape-compactness regularization in (6.11). The solution is derived from minimizing an overall objective integrating (6.11) and a unary term $\mathbf{u}^t\mathbf{v}$, which is based on the posterior probabilities of a discriminative

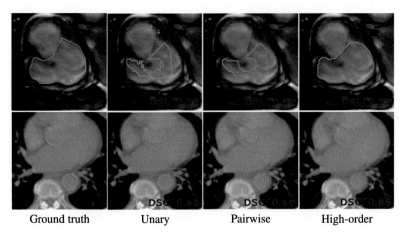

| Ground truth | Unary | Pairwise | High-order |

Figure 6.2 Examples of pairwise and high-order regularization. Segmentation of the right ventricle cavity in MRI (1st row) and the aorta in CT (2nd row). (*Figures from [21] © Springer.*)

convolutional neural network (CNN) classifier:

$$v_p = -\ln \frac{\Pr(1|\mathbf{f}_p, \boldsymbol{\theta})}{\Pr(0|\mathbf{f}_p, \boldsymbol{\theta})}.$$

In this case, $\boldsymbol{\theta}$ are the parameters of a CNN classifier, which are learned a priori from a large number of training images with ground-truth segmentations. The experimental examples in Fig. 6.2 illustrate the performance of shape-compactness regularization on two medical-imaging applications: segmenting the right ventricle cavity in Magnetic Resonance Imaging (MRI) and segmenting the aorta in computed tomography (CT) scans. In these examples, the unary potentials are obtained from training the deep 3D CNN architecture in [37], which includes nine convolutional layers with nonlinear activation units, three fully-connected layers (converted into standard convolution operations) and a soft-max layer. The fourth column in the figure shows how adding the high-order regularization in (6.11) improves the result obtained with the unary deep-network predictions.

Each displayed result includes the Dice similarity coefficient (DSC) as a performance measure.[1] Interestingly, standard pairwise MRF regularization

---

[1] DSC is a standard measure for evaluating medical image segmentation algorithms, which assesses a similarity between an obtained segmentation region $S \subset \Omega$ and a ground-truth segmentation region $G \subset \Omega$: DSC $= \frac{2|S \cap G|}{|S|+|G|}$, where $|.|$ denotes cardinality. The higher

might even decrease the performance of the CNN predictions, even though it yields smoother segmentation boundaries; see the aorta segmentations in the second row in the figure. This is due the well-known shrinking problem of standard MRF regularization.

6.3 Graph clustering

In the machine-learning community, ratio functions for clustering graph data are ubiquitous, e.g., Normalized Cut (NC) or Average Association (AA) functions [23,24], among others. Such graph clustering objectives are high-order functions. For instance, the popular NC objective for finding a partition $\mathbf{S} = \{S^l, l = 0, \ldots, L - 1\}$ of image domain Ω reads:

$$-\sum_{l=0}^{L-1} \frac{\sum_{p,q \in S^l} w_{pq}}{\sum_{p \in S^l} d_p} = -\sum_{l=0}^{L-1} \frac{(\mathbf{u}^l)^t W \mathbf{u}^l}{\mathbf{d}^t \mathbf{u}^l}, \qquad (6.13)$$

where, for each segment S^l, $\mathbf{u}^l = (u_{p,l})_{p \in \Omega}$ is a binary vector, which contains indicator variables of segment $S^l \subset \Omega$: $u_{p,l} = 1$ if pixel p belongs to S^l and $u_{p,l} = 0$ otherwise. $W = [w_{pq}]$ is an $|\Omega| \times |\Omega|$ matrix of pairwise potentials w_{pq}, each measuring a similarity between points p and q. For instance, the RBF kernel is a common choice as a pairwise similarity:

$$w_{pq} = \exp(-\|\mathbf{f}_p - \mathbf{f}_q\|^2/\alpha^2),$$

with α a hyper-parameter often referred to as kernel width. Vector $\mathbf{d} = (d_p)_{p \in \Omega}$ contains point degrees: $d_p = \sum_q w_{pq}$.

The recent work in [38] introduced Kernel Cut (KC), a single model that integrates a graph–clustering term and MRF regularization. The authors showed that KC could be powerful in the context of interactive segmentation because it integrates the benefits of graph clustering and MRFs. Indeed, high-order graph-clustering functions provide a potent alternative to standard model-fitting terms, e.g., probabilistic K-means objectives [25,26], which are widely used in conjunction with MRFs. Graph clustering uses pairwise similarities and, therefore, can deal effectively with high-dimensional features [23,38,39].

the DSC, the better the result. The measure takes its values in $[0, 1]$, with a value of 1 indicating a perfect match with the ground truth and a value of 0 indicating a total mismatch.

Model fitting has been successful when the image features are low-dimensional (e.g., $\mathbf{f}_p \in \mathbb{R}^3$ for color spaces) and are modeled with a low-complexity distribution within each segmentation region, such as the Gaussian distribution in the case of K-means. As pointed out in the previous chapter, deploying complex distributions such as histograms or Gaussian Mixture Models (GMMs) might be sensitive to local minima. It may result in over-fitting, a difficulty further increased with higher-dimensional feature spaces. This may explain why this class of likelihood-based, model-fitting clustering techniques are not common in the wider clustering community, and their use, although quite wide in the MRF community, has remained confined to segmenting low-dimensional image features.

Typically, MRFs, e.g., the Potts model, and graph-clustering objectives, such as NC or AA, are used separately in computer vision and machine-learning problems and applications. This might be explained by the significant differences in the applicable optimizers (spectral relaxation for NC and graph cuts or mean-field inference for MRFs). In the next chapter, we will discuss a general optimization solution [38], that enables the integration of graph-clustering objectives and MRFs.

Fig. 6.3 illustrates the benefit of using an NC clustering term instead of a model-fitting one for six-dimensional features, which include color-depth (RGBD) images along with the spatial coordinates. The first row shows the original images with bounding-box initialization, whereas the second depicts color-coded depth channels. The third row shows the results of the GrabCut algorithm (i.e., with a model-fitting data term), and the forth row shows the results of kernel cut (i.e., with a graph-clustering data term).

6.4 Distribution matching

An interesting problem consists of finding a foreground region $S = \{p \in \Omega \mid u_p = 1\}$ in one or several related images so that the distribution of some image features within S (e.g., color, textures, edge orientations, motion) most closely matches a given (learned) model distribution. Fig. 6.4 illustrates a use case for this problem. Each row depicts a set of related images containing the same target foreground region. The model distribution is learned from one single, manually segmented image. This is illustrated with the first image in each row in the figure, with the manual segmentation depicted with a green boundary. Then, given this model, related images are segmented automatically with an MRF-regularized distribution-matching function.

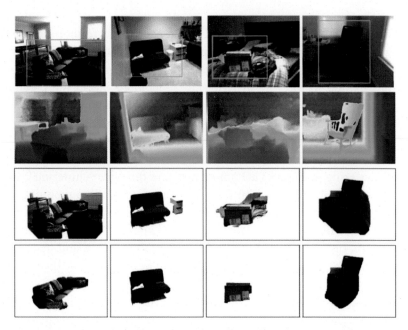

Figure 6.3 Interactive segmentation based on RGBD (color and depth) features and spatial coordinates using GrabCut and regularized graph clustering. First row: the original images with bounding-box initialization; second row: color-coded depth inputs; third row: GrabCut results; fourth row: results with MRF-regularized graph clustering. (*Figure from [38] © Springer.*)

This scenario occurs frequently in practice, for instance, in interactive segmentation of a video sequence or a set of personal smart-devise photos taken at the same scene. In these scenarios, it is practical to segment the same object in all related images, assuming a manual segmentation is provided for a single image. For instance, this could be very useful in data augmentation, increasing the sets of labeled images for supervised deep-learning based segmentation algorithms, thereby reducing the burden for creating dense pixel-wise labels.

Let $\mathbf{m} = \{m_{\mathbf{b}}, \mathbf{b} \in \mathcal{Z}\}$ denotes a probability simplex vector that represents the learned (known) distribution, where $\mathbf{b}$ denotes bins (e.g., color bins) belonging to a finite set $\mathcal{Z}$. For a new image containing the same target object (but potentially with variations in the object size and point of view), our purpose is to find a region $S = \{p \in \Omega \mid u_p = 1\}$ so that the distribution of features $\mathbf{f}_p$ in S most closely matches the known target distribution. To achieve this, one could use Kernel Density Estimation (KDE) along with

Figure 6.4 Distribution-matching examples.

the negative Bhattacharyya coefficient [9,13,40], among other possible distribution measures or divergences. Let $\mathbf{p} = \{p_{\mathbf{b}}, \mathbf{b} \in \mathcal{Z}\}$ denote the KDE estimate of the distribution of image features $\mathbf{f}_p \in \mathbb{R}^N$ within region S ($\mathbf{f}_p$ may represent color, for instance):

$$p_{\mathbf{b}}^{\mathbf{u}} = \frac{\sum_{p \in \Omega} u_p K_p^{\mathbf{b}}}{\sum_{p \in \Omega} u_p}, \tag{6.14}$$

where $K_p^{\mathbf{b}}$ a kernel function measuring a similarity between feature point $\mathbf{f}_p$ and bin $\mathbf{b}$, e.g., the Gaussian kernel:

$$K_p^{\mathbf{b}} = \frac{1}{(2\pi\sigma^2)^{\frac{N}{2}}} \exp^{-\frac{\|\mathbf{b}-\mathbf{f}_p\|^2}{2\sigma^2}}. \tag{6.15}$$

Parameter σ is the width of the kernel. One can also use the normalized histogram as density estimate, which corresponds to $K_p^{\mathbf{b}} = [\mathbf{f}_p = \mathbf{b}]$. Distribution matching [9,13,40] could be achieved by minimizing the negative

Bhattacharyya coefficient, which evaluates the overlap (similarity) between distributions $\mathbf{p}^u$ and learned model $\mathbf{m}$:

$$-\sum_{b\in Z} \sqrt{m_b p_b^u}. \tag{6.16}$$

The range of the negative Bhattacharyya coefficient is $[-1, 0]$, 0 corresponding to no overlap between the distributions and -1 to a perfect match. Fig. 6.4 depicts experimental examples that correspond to minimizing (6.16) along an MRF boundary regularization term.

References

[1] Y. Boykov, M.P. Jolly, Interactive graph cuts for optimal boundary and region segmentation of objects in n-d images, in: IEEE International Conference on Computer Vision (ICCV), 2001, pp. 105–112.

[2] Y. Boykov, G. Funka Lea, Graph cuts and efficient n-d image segmentation, International Journal of Computer Vision 70 (2) (2006) 109–131.

[3] Y. Boykov, O. Veksler, R. Zabih, Fast approximate energy minimization via graph cuts, IEEE Transactions on Pattern Analysis and Machine Intelligence 23 (11) (2001) 1222–1239.

[4] A. Blake, P. Kohli, C. Rother, Markov Random Fields for Vision and Image Processing, MIT Press, 2011.

[5] P. Krähenbühl, V. Koltun, Efficient inference in fully connected CRFs with Gaussian edge potentials, in: Advances in Neural Information Processing Systems (NIPS), 2011, pp. 109–117.

[6] Y. Boykov, V. Kolmogorov, An experimental comparison of min-cut/max-flow algorithms for energy minimization in vision, IEEE Transactions on Pattern Analysis and Machine Intelligence 26 (9) (2004) 1124–1137.

[7] M. Tang, D. Marin, I. Ben Ayed, Y. Boykov, Normalized cut meets MRF, in: European Conference on Computer Vision (ECCV), Part II, 2016, pp. 748–765.

[8] T. Taniai, Y. Matsushita, T. Naemura, Superdifferential cuts for binary energies, in: IEEE Conference on Computer Vision and Pattern Recognition (CVPR), 2015, pp. 2030–2038.

[9] I. Ben Ayed, K. Punithakumar, S. Li, Distribution matching with the Bhattacharyya similarity: A bound optimization framework, IEEE Transactions on Pattern Analysis and Machine Intelligence 37 (9) (2015) 1777–1791.

[10] Y. Boykov, H.N. Isack, C. Olsson, I. Ben Ayed, Volumetric bias in segmentation and reconstruction: Secrets and solutions, in: IEEE International Conference on Computer Vision (ICCV), 2015, pp. 1769–1777.

[11] Y. Kee, M. Souiai, D. Cremers, J. Kim, Sequential convex relaxation for mutual information-based unsupervised figure-ground segmentation, in: IEEE Conference on Computer Vision and Pattern Recognition (CVPR), 2014, pp. 4082–4089.

[12] Y. Lim, K. Jung, P. Kohli, Efficient energy minimization for enforcing label statistics, IEEE Transactions on Pattern Analysis and Machine Intelligence 36 (9) (2014) 1893–1899.

[13] L. Gorelick, F.R. Schmidt, Y. Boykov, Fast trust region for segmentation, in: IEEE Conference on Computer Vision and Pattern Recognition (CVPR), 2013, pp. 1714–1721.

[14] H. Jiang, Linear solution to scale invariant global figure ground separation, in: IEEE Conference on Computer Vision and Pattern Recognition (CVPR), 2012, pp. 678–685.

[15] V.Q. Pham, K. Takahashi, T. Naemura, Foreground-background segmentation using iterated distribution matching, in: IEEE International Conference on Computer Vision and Pattern Recognition (CVPR), 2011, pp. 2113–2120.

[16] H. Kervadec, J. Dolz, M. Tang, E. Granger, Y. Boykov, I. Ben Ayed, Constrained-CNN losses for weakly supervised segmentation, Medical Image Analysis 54 (2019) 88–99.

[17] M. Bateson, H. Lombaert, I. Ben Ayed, Test-time adaptation with shape moments for image segmentation, in: Medical Image Computing and Computer Assisted Intervention (MICCAI), in: LNCS, vol. 13434, 2022, pp. 736–745.

[18] M. Tang, A. Djelouah, F. Perazzi, Y. Boykov, C. Schroers, Normalized cut loss for weakly-supervised CNN segmentation, in: IEEE Conference on Computer Vision and Pattern Recognition (CVPR), 2018, pp. 1818–1827.

[19] H. Kervadec, H. Bahig, L. Letourneau-Guillon, J. Dolz, I. Ben Ayed, Beyond pixel-wise supervision: semantic segmentation with higher-order shape descriptors, in: Medical Imaging with Deep Learning (MIDL), 2021, pp. 1–16.

[20] M. Boudiaf, H. Kervadec, I.M. Ziko, P. Piantanida, I. Ben Ayed, J. Dolz, Few-shot segmentation without meta-learning: A good transductive inference is all you need?, in: IEEE Conference on Computer Vision and Pattern Recognition (CVPR), 2021, pp. 13979–13988.

[21] J. Dolz, I. Ben Ayed, C. Desrosiers, Unbiased shape compactness for segmentation, in: Medical Image Computing and Computer Assisted Intervention (MICCAI), Part I, 2017, pp. 755–763.

[22] M. Niethammer, C. Zach, Segmentation with area constraints, Medical Image Analysis 17 (1) (2013) 101–112.

[23] J. Shi, J. Malik, Normalized cuts and image segmentation, IEEE Transactions on Pattern Analysis and Machine Intelligence 22 (8) (2000) 888–905.

[24] U. Von Luxburg, A tutorial on spectral clustering, Statistics and Computing 17 (4) (2007) 395–416.

[25] C. Rother, V. Kolmogorov, A. Blake, GrabCut: Interactive foreground extraction using iterated graph cuts, ACM Transactions on Graphics 23 (3) (2004) 309–314.

[26] T. Chan, L. Vese, Active contours without edges, IEEE Transactions on Image Processing 10 (2) (2001) 266–277.

[27] M. Bateson, J. Dolz, H. Kervadec, H. Lombaert, I. Ben Ayed, Constrained domain adaptation for image segmentation, IEEE Transactions on Medical Imaging 40 (7) (2021) 1875–1887.

[28] F.S. He, Y. Liu, A.G. Schwing, J. Peng, Learning to play in a day: Faster deep reinforcement learning by optimality tightening, in: International Conference on Learning Representations (ICLR), 2017, pp. 1–13.

[29] D.P. Bertsekas, Nonlinear Programming, Athena Scientific, Belmont, MA, 1995.

[30] M. Klodt, D. Cremers, A convex framework for image segmentation with moment constraints, in: IEEE International Conference on Computer Vision (ICCV), 2011, pp. 2236–2243.

[31] A. Foulonneau, P. Charbonnier, F. Heitz, Multi-reference shape priors for active contours, International Journal of Computer Vision 81 (1) (2009) 68–81.

[32] P. Milanfar, M. Putinar, J. Varah, B. Gustafsson, G.H. Golub, Shape reconstruction from moments: Theory, algorithms, and applications, Advanced Signal Processing Algorithms, Architectures, and Implementations X, vol. 4116, vol. 4116, International Society for Optics and Photonics, 2000, pp. 406–416.

[33] M.K. Hu, Visual pattern recognition by moment invariants, IRE Transactions on Information Theory 8 (2) (1962) 179–187.

[34] V. Kolmogorov, Y. Boykov, C. Rother, Applications of parametric maxflow in computer vision, in: IEEE International Conference on Computer Vision (ICCV), 2007, pp. 1–8.

[35] L. Grady, E.L. Schwartz, Isoperimetric graph partitioning for image segmentation, IEEE Transactions on Pattern Analysis and Machine Intelligence 28 (3) (2006) 469–475.

[36] J.D. Zunic, K. Hirota, P.L. Rosin, A Hu moment invariant as a shape circularity measure, Pattern Recognition 43 (1) (2010) 47–57.

[37] J. Dolz, C. Desrosiers, I. Ben Ayed, 3D fully convolutional networks for subcortical segmentation in MRI: A large-scale study, NeuroImage 170 (2018) 456–470.

[38] M. Tang, D. Marin, I. Ben Ayed, D. Marin, Y. Boykov, Kernel cuts: Kernel & spectral clustering meet regularization, International Journal of Computer Vision 127 (5) (2019) 477–511.

[39] D. Marin, M. Tang, I. Ben Ayed, Y. Boykov, Kernel clustering: Density biases and solutions, IEEE Transactions on Pattern Analysis and Machine Intelligence 41 (1) (2018) 136–147.

[40] I. Ben Ayed, L. Gorelick, Y. Boykov, Auxiliary cuts for general classes of higher-order functionals, in: IEEE International Conference on Computer Vision and Pattern Recognition (CVPR), 2013, pp. 1304–1311.

CHAPTER 7

Pseudo-bound optimization

7.1 Bound optimization

As discussed in previous chapters, a large class of image segmentation techniques is based on minimizing regularization-based functionals of the following general form:

$$\mathcal{E}(\mathbf{S}) = \sum_{l=0}^{L-1} \mathcal{J}(S^l) + \lambda \mathcal{R}(\mathbf{S}), \tag{7.1}$$

where $\mathbf{S} = \{S^l, l = 0, \dots, L-1\}$ is a set of variables denoting L segmentation regions $S^l \subset \Omega$, with $\Omega \subset \mathbb{R}^2$ the spatial image domain. $\mathcal{R}(\mathbf{S})$ is a regularization term evaluating the boundary length of segmentation $\mathbf{S}$ via Euclidean or some edge-sensitive, feature-weighted metric, and λ is a positive constant balancing the effect of this term. We examined several examples where each $\mathcal{J}(S^l)$ is a high-order regional term, and we illustrated the usefulness of such terms in practice. Unfortunately, powerful and global optimization techniques, such as graph cuts [1,2], are restricted to special forms of regional and boundary functions. For instance, regional terms are typically linear (or unary) in the context of graph-cut segmentation, i.e., for a given region S, $\mathcal{J}(S)$ takes the form of individual-pixel penalties [3,4]:

$$\mathcal{J}(S) = \sum_{p \in S} v_p = \sum_{p \in \Omega} u_p v_p = \mathbf{u}^t \mathbf{v}, \tag{7.2}$$

where[1]:

- $\mathbf{u} = (u_p)_{p \in \Omega} \in \{0, 1\}^{|\Omega|}$ is a binary vector that contains the indicator variables of segment $S \subset \Omega$: $u_p = 1$ if pixel p belongs to S and $u_p = 0$ otherwise.

---

[1] In a large part of this chapter, and for the sake of presentation clarity, the bound and pseudo-bound description might refer to a single region S, i.e., we omit superscript l for the notation of the regions and the corresponding indicator variables. This will be clear from the context and will not lead to any ambiguity.

High-Order Models in Semantic Image Segmentation
https://doi.org/10.1016/B978-0-12-805320-1.00012-9

- $\mathbf{v} = (v_p)_{p\in\Omega} \in \mathbb{R}^{|\Omega|}$ is a vector containing a unary potential v_p for each p. The regional term of popular Boykov–Jolly model [3,4], which we discussed in Chapter 1, corresponds to the linear form (7.2) with v_p given by:

$$v_p = -\ln \frac{\Pr(\mathbf{f}_p|1,\boldsymbol{\theta})}{\Pr(\mathbf{f}_p|0,\boldsymbol{\theta})}, \tag{7.3}$$

where $\Pr(\mathbf{f}_p|l,\boldsymbol{\theta})$ are fixed probability models for the foreground ($l = 1$) and background ($l = 0$) regions. $\mathbf{f}_p \in \mathbb{R}^N$, $p \in \Omega$, denote image features, e.g., color, intensity or textures, and $\boldsymbol{\theta}$ is a set of parameters. We also discussed examples of boundary regularization terms $\mathcal{R}(\mathbf{S})$, which are typically written in terms of pairwise potentials. In particular, we emphasized the Potts model [1,5,6], an important example of pairwise regularization and a mainstay in computer vision. Recall that Potts regularization belongs to the family of submodular functions, which were instrumental in the development of various efficient computer vision algorithms. For instance, the global optimum of a function containing unary and submodular pairwise potentials can be computed exactly in polynomial time (with respect to image size) using graph cuts [2]. As they are easily amenable to powerful and global optimizers, such unary (first-order) and submodular pairwise (second-order) terms are massively used in computer vision and medical image analysis. Unfortunately, the set of properties that these terms can impose on the obtained solutions are limited. Therefore, high-order [7–15] and nonsubmodular [8,16–18] functions have attracted substantial research efforts in the field. As discussed in Chapter 6, such difficult-to-optimize terms occur frequently in a breadth of computer vision, image processing and data analysis tasks. For instance, fractional terms are prominent in the context of graph clustering, e.g., the very popular normalized cut objective [19,20]. High-order functions can impose various priors on image segmentation. This includes high-order regularization [21], priors on the distributions/histograms of features within the target regions [9] or priors on segment size/shape [13], which can be very useful in medical-imaging applications [22,23]. Nonsubmodular terms arise in problems such as curvature regularization [24], surface registration [16], deconvolution and inpainting [17], among others. This chapter examines a class of iterative techniques for optimizing high-order and nonsubmodular discrete functions using bounds or pseudo-bounds [25].

7.2 Bound optimization

Bound optimizers iteratively minimize an auxiliary function that is a tight upper bound of the original objective function.

Definition 1 (Auxiliary function). $\mathcal{A}_i(S)$ is an auxiliary function of $\mathcal{J}(S)$ at current solution S_i, if it satisfies the following conditions:

$$\mathcal{J}(S) \leq \mathcal{A}_i(S), \; \forall S \tag{7.4a}$$

$$\mathcal{J}(S_i) = \mathcal{A}_i(S_i). \tag{7.4b}$$

In these conditions, i denotes the iteration counter. Now, we have to update current solution S_i to the optimum of the auxiliary function:

$$S_{i+1} = \arg\min_{S} \mathcal{A}_i(S), \quad i = 1, 2, \ldots. \tag{7.5}$$

Ideally, optimizing the auxiliary function is easier than the original problem. Bound optimizers guarantee to not increase the original objective function at each iteration:

$$\mathcal{J}(S_{i+1}) \leq \mathcal{A}_i(S_{i+1}) \leq \mathcal{A}_i(S_i) = \mathcal{J}(S_i). \tag{7.6}$$

Inequality (7.6) can be obtained easily from the auxiliary-function conditions in (7.4) and from the fact that S_{i+1} is the minimum of $\mathcal{A}_i(S)$, which implies $\mathcal{A}_i(S_{i+1}) \leq \mathcal{A}_i(S_i)$. Fig. 7.1 illustrates the general principle of bound optimization.

Bound optimization techniques replace a difficult high-order problem by a sequence of easier problems [26]. Examples of well-known bound optimizers include mean-shift [27], concave-convex procedures [28], expectation maximization (EM) and submodular–supermodular procedures [29]. Various problems were effectively addressed with bound optimization, in areas such as machine learning [26], computational statistics [30] and nonnegative matrix factorization [31]. In the context of discrete high-order or nonsubmodular models, several recent computer vision works showed that bound optimizers can be very useful in practice [7], and can yield competitive optimization performances, e.g., [8,25,32], among many other works. Furthermore, such procedures are parameter- and derivative-free. In fact, they do not depend on optimization parameters such as step sizes, neither do they require the function to be differentiable. This makes them very convenient for a breadth of difficult discrete problems in computer vision. The main difficulty in bound optimization is in building an

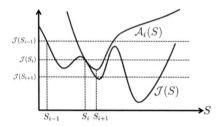

Figure 7.1 Bound optimization. The auxiliary function $\mathcal{A}_i$ is depicted by red color and the original objective function $\mathcal{J}(S)$ by blue. Minimizing the auxiliary function $\mathcal{A}_i$ at iteration $(i+1)$ yields a solution S_{i+1}, which guarantees the original function does not increase: $\mathcal{J}(S_{i+1}) \leq \mathcal{J}(S_i)$.

appropriate auxiliary function. On the one hand, a good bound should approximate well the original function over the space of possible solutions. On the other hand, such a bound should be amenable to global and fast optimizers such as graph cuts.

7.2.1 Examples of bound optimizers

In the following, we will discuss bound–optimization perspectives for several algorithms based on regularization, including the well-known GrabCut [33] and MRF-regularized graph clustering [7].

7.2.1.1 K-means and it probabilistic generalizations

In Chapter 4, we examined the K-means clustering objective and its probabilistic generalizations. We saw that a broad class of segmentation algorithms integrate probabilistic K-means (or model fitting) terms and regularization, taking the form (7.1), with $\mathcal{J}(S)$ given by[2]:

$$\mathcal{J}(S) = -\sum_{p \in S} \ln \Pr(\mathbf{f}_p | \boldsymbol{\theta}_S). \tag{7.7}$$

A term of this form is high–order due to the dependence of segment S on its set of parameters $\boldsymbol{\theta}_S$. A standard way to optimize a sum of terms of the form (7.7) is to alternate two steps until convergence: one finding the optimal parameters while the regions are fixed and the other optimizing

---

[2] Notice that, here, we changed slightly the notation of the likelihood in Chapter 4. We omitted the index of the region and made the set of parameters dependent on variable S, which will be convenient in subsequent bound-optimization discussions.

with respect to the segmentation while the parameters are fixed. The basic K-means procedure and popular regularization-based segmentation algorithms, such as GrabCut [33] and the Chan–Vese model [34], belong to this class of block-coordinate descent procedures, which guarantee that the objective function does not increase at every step. In fact, such standard procedures can be viewed as bound optimizers [25,35]. Let us come back to the basic K-means procedure, which we discussed in Chapter 4. Recall that the K-means objective function, whose optimization is NP-hard [35], can be written as follows:

$$\sum_{l=0}^{L-1}\sum_{p\in S^l}\|\mathbf{f}_p - \mathbf{m}_{S^l}\|^2 \text{ where } \mathbf{m}_{S^l} = \frac{1}{|S^l|}\sum_{p\in S^l}\mathbf{f}_p. \tag{7.8}$$

Expressing $\mathbf{m}_{S^l}$ as a function of variable S^l results in high-order terms that depend solely on the segmentation.

The standard K-means procedure alternates two steps until convergence. At iteration i, and given segments S_i^l obtained from the previous iteration, the first step updates the mean of the features within each region S_i^l:

$$\textbf{Step I} \text{ (parameter update): } \mathbf{m}_{S_i^l} = \frac{1}{|S_i^l|}\sum_{p\in S_i^l}\mathbf{f}_p. \tag{7.9}$$

Now, given fixed parameters $\mathbf{m}_{S_i^l}$, the second step consists of optimizing the objective in (7.8) with respect to all segments S^l:

Step II (partition update):

$$\{S_{i+1}^l, \, l=0,\ldots,L-1\} = \underset{\{S^l, \, l=0,\ldots,L-1\}}{\arg\min} \sum_{l=0}^{L-1}\sum_{p\in S^l}\|\mathbf{f}_p - \mathbf{m}_{S_i^l}\|^2. \tag{7.10}$$

With the region means fixed, (7.10) is a sum of unary potentials. Therefore, the global optimum with respect to $\{S^l, \, l=0,\ldots,L-1\}$ can be obtained trivially by assigning each feature point to the region having the closest mean:

$$p \text{ is assigned to } S_{i+1}^l \text{ if } \|\mathbf{f}_p - \mathbf{m}_{S_i^l}\| \le \|\mathbf{f}_p - \mathbf{m}_{S_i^k}\| \, \forall k \ne l. \tag{7.11}$$

Proposition 1. *The iterative K-means updates in (7.9) and (7.11) correspond to a bound optimization for high-order term (7.8). Given* $\{S_i^l, \, l=0,\ldots,L-1\}$

defining the solution at the current iteration, the auxiliary function is:

$$\mathcal{A}_i(\{S^l, \, l = 0, \ldots, L-1\}) = \sum_{l=0}^{L-1} \sum_{p \in S^l} \|\mathbf{f}_p - \mathbf{m}_{S_i^l}\|^2. \tag{7.12}$$

Proof. Consider the derivative of function $\sum_{p \in S^l} \|\mathbf{f}_p - \mathbf{y}\|^2$ with respect to $\mathbf{y}$. Setting this derivative equal to zero yields a closed-form solution, which corresponds to the mean of features within S^l: $\tilde{\mathbf{y}} = \mathbf{m}_{S^l} = \frac{1}{|S^l|} \sum_{p \in S^l} \mathbf{f}_p$. Notice that $\sum_{p \in S^l} \|\mathbf{f}_p - \mathbf{y}\|^2$ is convex with respect $\mathbf{y}$, which implies $\mathbf{m}_{S^l}$ is its global minimum. Therefore, we have the following inequality, which gives a unary upper bound on each term of the K-means objective:

$$\sum_{p \in S^l} \|\mathbf{f}_p - \mathbf{m}_{S^l}\|^2 \le \sum_{p \in S^l} \|\mathbf{f}_p - \mathbf{m}_{S_i^l}\|^2. \tag{7.13}$$

Summing (7.13) over the segments, it follows directly that $\mathcal{A}_i$ in (7.12) is an upper bound on the high-order K-means objective in (7.8):

$$\sum_{l=0}^{L-1} \sum_{p \in S^l} \|\mathbf{f}_p - \mathbf{m}_{S^l}\|^2 \le \mathcal{A}_i(\{S^l, \, l = 0, \ldots, L-1\}). \tag{7.14}$$

From expression (7.12), it is easy to verify that $\mathcal{A}_i$ is a tight upper bound at current solution $\{S_i^l, \, l = 0, \ldots, L-1\}$, i.e., we have equality in (7.14) when $S^l = S_i^l \, \forall l$. Therefore, $\mathcal{A}_i$ is an auxiliary function of K-means. $\square$

In summary, step I (parameter update) computes the means of features within the regions, $\{\mathbf{m}_{S_i^l}, \, l = 0, \ldots, L-1\}$, yielding auxiliary function $\mathcal{A}_i$. Step 2 (partition update) optimizes the bound with respect to partition variables $\{S^l, \, l = 0, \ldots, L-1\}$. Fig. 7.2 illustrates this bound optimization view of K-means.

This bound-optimization perspective of K-means can be extended to the probabilistic generalization in (7.7), which we examined in great detail earlier in the book. Recall that, using Monte Carlo estimation, we can show that the likelihood term of region S, which takes the form (7.7), is an approximation of the region entropy multiplied by region size (or cardinality):

$$|S|\mathcal{H}(\mathbf{f}|\boldsymbol{\theta}_S) \approx -\sum_{p \in S} \ln \Pr(\mathbf{f}_p|\boldsymbol{\theta}_S). \tag{7.15}$$

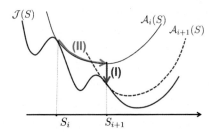

Figure 7.2 A bound optimization perspective of the K-means procedure. Step I (parameter update) computes the means of features within the regions, yielding next auxiliary function $\mathcal{A}_{i+1}$. Step 2 (partition update) optimizes current auxiliary function $\mathcal{A}_i$ with respect to partition variables. (*Figure adopted from [35] © Springer.*)

The approximation is exact (i.e., equality holds) when probability models $\Pr(\mathbf{f}_p|\theta_S)$ are histograms. From (7.15), it follows that the popular GrabCut algorithm [33] can be viewed as an optimizer for the high-order entropy function in the case of color histograms, integrated with a boundary-regularization term:

$$\sum_{l=0}^{L-1} |S^l| \mathcal{H}(\mathbf{f}|\theta_{S^l}) + \lambda \mathcal{R}(\mathbf{S}). \tag{7.16}$$

More generally, due to approximation (7.15), all standard model-fitting techniques, which follow alternating schemes for optimizing general likelihood terms (7.7), can be viewed as approximate and indirect optimizers of the high-order entropy in (7.16). In the following, we will see that the alternating-optimization scheme of GrabCut [33] can be viewed as a bound optimizer when using color histograms. This perspective of GrabCut, discussed in [25], allows building more efficient optimizers for the entropy objective via *pseudo-bounds*, which we will discuss in greater detail later.

Proposition 2. *GrabCut is a bound optimizer for regularized high-order entropy objective* (7.16) *when probability models* θ_{S^l} *correspond to normalized histograms. Let us assume that we have color histograms* $\theta_{S_i^l}$, $l = 0, \ldots, L-1$ *that are computed from regions* S_i^l *obtained at the previous iteration i. Then, we have the following auxiliary function for problem* (7.16):

$$\mathcal{A}_i(\{S^l, l=0,\ldots,L-1\}) = -\sum_{p \in S^l} \ln \Pr(\mathbf{f}_p|\theta_{S_i^l}) + \lambda \mathcal{R}(\mathbf{S}). \tag{7.17}$$

Notice that the auxiliary function in (7.17) is exactly the sum of unary and regularization terms that GrabCut optimizes at each iteration via a graph cut.

Proof. The proof follows directly from a cross-entropy argument [25]. In fact, using Monte Carlo estimation, we can easily show the following:

$$|S^l|\mathcal{H}(\mathbf{f}|\boldsymbol{\theta}_{S^l}, \boldsymbol{\theta}_{S_i^l}) \approx -\sum_{p \in S^l} \ln \Pr(\mathbf{f}_p|\boldsymbol{\theta}_{S_i^l}), \tag{7.18}$$

where $\mathcal{H}(\mathbf{f}|\boldsymbol{\theta}_{S^l}, \boldsymbol{\theta}_{S_i^l})$ is the cross-entropy between $\Pr(\mathbf{f}_p|\boldsymbol{\theta}_{S^l})$, the distribution of features within variable region S^l, and $\Pr(\mathbf{f}_p|\boldsymbol{\theta}_{S_i^l})$, the distribution of features within current region S_i^l. The approximation is exact in the case of histograms. Now, it is well known that the cross-entropy between two distributions A and B verifies the following: $\mathcal{H}(A, B) \geq \mathcal{H}(A, A) = \mathcal{H}(A)$. Therefore,

$$\mathcal{H}(\mathbf{f}|\boldsymbol{\theta}_{S^l}, \boldsymbol{\theta}_{S_i^l}) \geq \mathcal{H}(\mathbf{f}|\boldsymbol{\theta}_{S^l}). \tag{7.19}$$

The fact that (7.17) is an upper bound on high-order entropy functional (7.16) follows directly from (7.18) and (7.19). It is easy to verify that the bound is tight at the current solution $\mathcal{H}(\mathbf{f}|\boldsymbol{\theta}_{S_i^l}, \boldsymbol{\theta}_{S_i^l}) = \mathcal{H}(\mathbf{f}|\boldsymbol{\theta}_{S_i^l})$.   □

Recall that GrabCut iterates two steps, one performing a graph-cut segmentation with fixed color models, as in the case of the Boykov–Jolly model [3], and the other updating color models with the segmentation fixed. Proposition 2 means that the graph-cut step corresponds to optimizing the auxiliary function, yielding the next segmentation, whereas the model updates compute the next auxiliary function.

7.2.1.2 Ratio functionals for graph clustering

In Chapter 6, we discussed very popular ratio functions for graph clustering, e.g., normalized cut (NC) and average association (AA). Such high-order functions take the following general form for finding a partition $\mathbf{S} = \{S^l, l = 0, \ldots, L-1\}$ of image domain Ω:

$$-\sum_{l=0}^{L-1} \frac{\sum_{p,q \in S^l} K_{pq}}{\sum_{p \in S^l} d_p} = -\sum_{l=0}^{L-1} \frac{(\mathbf{u}^l)^t \mathcal{K} \mathbf{u}^l}{\mathbf{d}^t \mathbf{u}^l}, \tag{7.20}$$

where, for each segment S^l, $\mathbf{u}^l = (u_{p,l})_{p \in \Omega}$ is a binary vector that contains indicator variables of segment $S^l \subset \Omega$: $u_{p,l} = 1$ if pixel p belongs to S^l and

$u_{p,l} = 0$ otherwise. Recall that $\mathcal{K} = [K_{pq}]$ is an $|\Omega| \times |\Omega|$ matrix of pairwise potentials K_{pq}, each evaluating an affinity (similarity) between points p and q, e.g., via some kernel function. d_p denotes the degree of point p: $d_p = \sum_q K_{pq}$. Within the learning community, ratio functionals of the general form (7.20) are widely used for partitioning high-dimensional features [35, 36], and standard spectral relaxation [19,20] is a dominant technique for optimizing such functionals.

Moreover, Chapter 6 discussed segmentation examples that demonstrated that integrating graph-clustering terms and MRF regularizers in a single model can be powerful, for instance:

$$\mathcal{E}(\mathbf{S}) = -\sum_{l=0}^{L-1} \frac{\sum_{p,q \in S^l} K_{pq}}{\sum_{p \in S^l} d_p} + \lambda \mathcal{R}(\mathbf{S}), \tag{7.21}$$

where $\mathcal{R}(\mathbf{S})$ is some standard edge-alignment and boundary-regularization term such as Potts. MRFs and graph clustering have complementary advantages and limitations.

One the one hand, graph clustering uses pairwise similarities and, therefore, can deal effectively with high-dimensional features [19,35,36]. However, standard MRF regularizers are typically used in conjunction with model-fitting terms, e.g., probabilistic K-means objectives [33,34]. In general, model fitting is appropriate when the features follow a low–complexity model within each segmentation region, e.g., a Gaussian distribution, as is the case of K-means. As pointed out in several works [37,38], using complex models such as histograms of Gaussian mixture models (GMMs) is sensitive to local minima and might result in over-fitting, even in the context of low-dimensional feature spaces such as color. This sensitivity to local minima might be further accrued in the case of higher-dimensional feature spaces, which might explain why this class of model-fitting clustering techniques is not common beyond the context of segmenting color images ($\mathbf{f}_p \in \mathbb{R}^3$). Therefore, high-order graph-clustering functions of the form in (7.21) provide a potent alternative to standard model-fitting terms. On the other hand, graph clustering applications can be enhanced by imposing additional constraints on the solutions [35,39]. For instance, it is well known that segmentation with the standard NC objective yields poor alignment to high-contrast edges, a problem that can be handled by adding a standard Potts regularization [35].

Unfortunately, integrating terms/constraints and graph-clustering functions, as in Eq. (7.21), typically results in optimization problems that are not amenable to spectral relaxation, the dominant technique for dealing with ratio functionals of the general form (7.20). Typically, MRFs, e.g., Potts, and graph-clustering objectives, such as NC or AA, are used separately in computer vision and machine-learning problems. This might be explained by the significant differences in the applicable optimizers. The work in [35] proposed a general bound-optimization solution that makes integrating graph-clustering objectives of the general form (7.20) and a wide class of MRFs amenable to powerful graph-cut optimizers such as expansion and swap moves [1]. In this general framework, graph-clustering functionals are viewed as high-order terms that can be replaced by unary auxiliary functions as discussed in the following.

Proposition 3 (Concavity of graph-clustering functionals). *Let* $\mathcal{J} : \mathbb{R}^{|\Omega|} \to \mathbb{R}$ *denote the following ratio functional:*

$$\mathcal{J}(\mathbf{u}) = -\frac{\mathbf{u}^t \mathcal{K} \mathbf{u}}{\mathbf{d}^t \mathbf{u}}. \tag{7.22}$$

When symmetric affinity matrix $\mathcal{K}$ is a positive semi-definite (PSD), ratio functional $\mathcal{J}(\mathbf{u})$ in (7.22) is concave over region $\mathbf{d}^t\mathbf{u} > 0$.

Proof. The result follows directly from the Hessian of $\mathcal{J}(\mathbf{u})$, which can be expressed as follows for symmetric matrix $\mathcal{K}$ [35]:

$$\frac{\nabla\nabla\mathcal{J}}{2} = -\frac{\mathcal{K}}{\mathbf{d}^t\mathbf{u}} + \frac{\mathcal{K}\mathbf{u}\mathbf{d}^t + \mathbf{d}\mathbf{u}^t\mathcal{K}}{(\mathbf{d}^t\mathbf{u})^2} - \frac{\mathbf{d}\mathbf{u}^t\mathcal{K}\mathbf{u}\mathbf{d}^t}{(\mathbf{d}^t\mathbf{u})^3}$$
$$\equiv -\frac{1}{\mathbf{d}^t\mathbf{u}}\left(\mathbf{I} - \frac{\mathbf{u}\mathbf{d}^t}{\mathbf{d}^t\mathbf{u}}\right)^t \mathcal{K}\left(\mathbf{I} - \frac{\mathbf{u}\mathbf{d}^t}{\mathbf{d}^t\mathbf{u}}\right).$$

It is easy to check that the Hessian is negative semi-definite (NSD) for PSD affinity matrix $\mathcal{K}$ and for variables $\mathbf{u}$ within region $\mathbf{d}^t\mathbf{u} > 0$.   □

For a given concave function $\mathcal{J}(\mathbf{u})$, the first-order approximation at current solution $\mathbf{u}_i$, with i the iteration counter, is an auxiliary function[3] of $\mathcal{J}$:

$$\mathcal{J}(\mathbf{u}) \leq \mathcal{J}(\mathbf{u}_i) + \nabla\mathcal{J}(\mathbf{u}_i)^t (\mathbf{u} - \mathbf{u}_i), \tag{7.23}$$

---

[3] Concavity arguments are commonly used in bound-optimization algorithms for deriving auxiliary functions [30]. In fact, well-known concave–convex procedures (CCCP) [28] are a particular case of this. Given a function expressed as the sum of concave and convex functions, one replaces the concave part by its first-order approximation, which results in a convex auxiliary function of the overall objective.

where $\nabla \mathcal{J}$ is the gradient of $\mathcal{J}$ that is given by:

$$\nabla \mathcal{J}(\mathbf{u}_i) = \mathbf{d} \frac{\mathbf{u}_i{}^t \mathcal{K} \mathbf{u}_i}{(\mathbf{d}^t \mathbf{u}_i)^2} - \mathcal{K} \mathbf{u}_i \frac{2}{\mathbf{d}^t \mathbf{u}_i}. \tag{7.24}$$

Eq. (7.23) implies that, at current solution $\mathbf{u}_i$, we have the following unary (linear) auxiliary function for $\mathcal{J}(\mathbf{u})$ in (7.22), up to an additive constant:

$$\mathcal{A}_i(\mathbf{u}) \overset{c}{=} \nabla \mathcal{J}(\mathbf{u}_i)^t \mathbf{u} = \sum_{p \in S} \nabla \mathcal{J}_p(\mathbf{u}_i), \tag{7.25}$$

where $S = \{p | u_p = 1\}$, $\nabla \mathcal{J}_p(\mathbf{u}_i)$ is component p of gradient vector $\nabla \mathcal{J}(\mathbf{u}_i) \in \mathbb{R}^{|\Omega|}$ and symbol $\overset{c}{=}$ denotes equality up to an additive constant. From this result, it is easy to see that the following is an auxiliary function of regularized graph-clustering functional (7.21), up to an additive constant:

$$-\sum_{l=0}^{L-1} \sum_{p \in S^l} \nabla \mathcal{J}_p(\mathbf{u}_i^l) + \lambda \mathcal{R}(\mathbf{S}), \tag{7.26}$$

with $\mathbf{u}_i^l$ the binary indicator vector of S_i^l, which denotes a segment obtained at iteration i. For a wide class of MRF regularizers $\mathcal{R}(\mathbf{S})$, bound (7.26) is amenable to powerful graph-cut optimizers such as expansion and swap moves [1] since the first term in (7.26) is unary. This provides a general bound-optimization solution that makes possible integrating graph clustering and MRFs.

The bound in (7.26) is based on the concavity of $\mathcal{J}$, which requires matrix $\mathcal{K}$ to be PSD. To generalize the result for arbitrary $\mathcal{K}$, one can use a diagonal shift [35], replacing $\mathcal{K}$ by $\tilde{\mathcal{K}} = \delta D + A$, where D is a degree matrix whose diagonal elements are given by d_p: $D = \text{diag}(\mathbf{d})$. For binary variables, it easy to check that this diagonal shift adds a constant to ratio functional (7.22):

$$\tilde{\mathcal{J}}(\mathbf{u}) = -\frac{\mathbf{u}^t \tilde{\mathcal{K}} \mathbf{u}}{\mathbf{d}^t \mathbf{u}} = -\frac{\mathbf{u}^t (\delta D + \mathcal{K}) \mathbf{u}}{\mathbf{d}^t \mathbf{u}} = -\delta - \frac{\mathbf{u}^t \mathcal{K} \mathbf{u}}{\mathbf{d}^t \mathbf{u}} \overset{c}{=} \mathcal{J}(\mathbf{u}). \tag{7.27}$$

This is due to the fact that, for binary vectors $\mathbf{u} \in \{0, 1\}^{|\Omega|}$, we have:

$$\mathbf{u}^t D \mathbf{u} = \mathbf{u}^t \text{diag}(\mathbf{d}) \mathbf{u} = \mathbf{d}^t \mathbf{u}.$$

Notice that $\tilde{\mathcal{K}} = \delta D + \mathcal{K}$ is PSD for sufficiently large δ when $d_p > 0 \, \forall p \in \Omega$. Proposition 3 means that the first-order approximation of $\tilde{\mathcal{J}}$ is an auxiliary function for $\tilde{\mathcal{J}}$. Given the equivalence between $\tilde{\mathcal{J}}$ and $\mathcal{J}$ for binary variables, one can use the auxiliary function of $\tilde{\mathcal{J}}$ to optimize $\mathcal{J}$. This guarantees that $\mathcal{J}$ does not increases at each iteration even when $\mathcal{K}$ is not PSD.

7.3 Pseudo-bound optimization

Standard bound optimization typically constructs a single auxiliary function (upper bound) and optimizes it at each iteration. In this case, the auxiliary function should be an upper bound of functional $\mathcal{J}(S)$ over the space of all possible solutions; see the function depicted in red in Fig. 7.1. In practice, it is difficult to obtain auxiliary functions that are (i) good approximations of the initial functional and (ii) amenable to global and efficient optimization techniques. As discussed earlier, bound optimization can be efficient in many computer vision applications [7,8,17]. However, there are several instances of high-order functions where optimizing a single auxiliary function yields weak solutions. The distribution matching example in Fig. 7.6, which we will discuss in greater detail later in this chapter, illustrates this.

The idea of pseudo-bound optimization is to relax the condition that an auxiliary function should be an upper bound on $\mathcal{J}$ over all possible solutions [25], which might yield better approximations of the original functional. Fig. 7.3 depicts an illustration of the concept. The red curve is an auxiliary function $\mathcal{A}_i(S)$, which is larger than $\mathcal{J}(S)$ at any solution S. The global minimum of $\mathcal{A}_i(S)$ guarantees that $\mathcal{J}(S)$ does not increase at each iteration. However, one can find many other approximations of $\mathcal{J}(S)$, with the global minima of such approximations also decreasing the original functional. Consider, for instance, the functions corresponding to the green curves (a) and (c) in Fig. 7.3, which we refer to as pseudo-bounds [25] because they do not dominate $\mathcal{J}(S)$ over all possible solutions S. Locally, around their global minima, these functions are upper bounds on $\mathcal{J}(S)$. Therefore, minimizing each of these functions decreases $\mathcal{J}(S)$ at the current iteration. Function (b) is not a local upper bound of $\mathcal{J}(S)$ in the neighborhood of its global minimum. However, its minimization still decreases $\mathcal{J}(S)$ at the current iteration. Notice that, at the current iteration i, the global minimum of function (c) yields a much better decrease of original functional $\mathcal{J}(S)$ than what one would have obtained with auxiliary function $\mathcal{A}_i(S)$. Therefore, we can obtain better solutions if we can explore efficiently all the minima of the pseudo-bounds (a, b and c). The best among these might yield a better optimum than the auxiliary function $\mathcal{A}_i$.

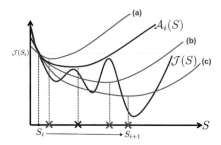

Figure 7.3 Pseudo-bound optimization. The auxiliary function $\mathcal{A}_i$ is depicted in red, the pseudo-bounds in green and the original objective function $\mathcal{J}(S)$ in blue. (*Figure adopted from [25] © Springer.*)

The main question is how to design pseudo-bounds in a way that guarantees the original functional does not increase at each iteration, without compromising significantly computational efficiency. Recall from (7.6) that, for an auxiliary function, the upper-bound condition was necessary to prove that $\mathcal{J}(S)$ does not increase at each iteration. One possible solution is to design some specific pseudo-bounds, which, by construction, guarantee the original functional does not increase. We can have such a guarantee if, for instance, we manage to find a pseudo-bound that dominate $\mathcal{J}(S)$ locally, in the vicinity of its global minimum, as is the case of functions (a) and (c) in Fig. 7.3.

An alternative approach would be to optimize, at each iteration, a family of pseudo-bound functions that includes at least one auxiliary function, i.e., the upper-bound constraint does not have to be satisfied for all the functions. In fact, if we can select efficiently the best solution among all the minima of the family, then we can guarantee that the original functional does not increase at each iteration. This guarantee comes from the fact that the family includes an auxiliary function. In the remainder of this chapter, we will discuss this more formally, and show how to construct pseudo-bounds for a large class of applications and problems. We will also include experimental results confirming that pseudo-bounds improve significantly the quality of the solutions obtained with a single auxiliary function.

7.3.1 Parametric pseudo-bounds

In the following, we discuss one parametric way to build a family of pseudo-bounds for which we can explore all minima efficiently [25]. For a given high-order function, we start by finding a submodular auxiliary function that can be optimized with powerful techniques such as graph cuts. Later in this chapter, we will discuss ways of obtaining such auxiliary functions for a general class of high-order terms. Then, we augment the auxiliary function with a unary-potential term multiplied by a parameter α. This relaxes the bound constraint for $\alpha \neq 0$.

Parametric max-flow algorithms [40] can compute efficiently a finite set of all distinct solutions $\mathbf{u}(\alpha) \in \{0, 1\}^{|\Omega|}$ of a parametrized family of discrete problems having the general form:

$$\mathbf{u}(\alpha) = \underset{\mathbf{u} \in \{0,1\}^{|\Omega|}}{\arg\min} \overbrace{\sum_{p \in \Omega} (a_p + \alpha b_p) u_p}^{\mathcal{A}(\mathbf{u}, \alpha)} + \sum_{(p,q) \in \mathcal{N}} \psi_{p,q}(u_p, u_q), \qquad (7.28)$$

where $\psi_{p,q}$ are pairwise submodular functions for a neighborhood system $\mathcal{N}$. The unary potentials in (7.28) are linear functions of α. Furthermore, they should be monotone with respect to α (i.e., coefficients b_p are either all positive or all negative) in order for parametric max-flow to find all the global minima of the parametric family in low-order polynomial time [25, 40]. This condition is important in practice when choosing a parametric family of pseudo-bounds; we will examine practical examples later in this chapter. Let us first discuss pseudo-bound family formally, and consider the following definition for a scalar parameter $\alpha \in [\alpha_{\min}, \alpha_{\min}] \subseteq \mathbb{R}$.

Definition 2 (Pseudo-bound). Let $\mathcal{J}(\mathbf{u})$ denotes a high-order function, $\mathbf{u}_i \in \{0, 1\}^{|\Omega|}$ the current solution at iteration i and $\alpha \in [\alpha_{\min}, \alpha_{\min}] \subseteq \mathbb{R}$ a parameter. We call parametric family of functions $\mathcal{F}_i(\mathbf{u}, \alpha) : \{0, 1\}^{|\Omega|} \times [\alpha_{\min}, \alpha_{\min}] \to \mathbb{R}$ a pseudo-bound for $\mathcal{J}(\mathbf{u})$ if there exists a least one value $\alpha' \in [\alpha_{\min}, \alpha_{\min}]$ verifying $\mathcal{F}_i(\mathbf{u}, \alpha')$ is an auxiliary function for $\mathcal{J}(\mathbf{u})$ at current solution $\mathbf{u}_i$.

The purpose is to update iteratively $\mathbf{u}_i$ for decreasing functional $\mathcal{J}(\mathbf{u})$ at each iteration i. Therefore, instead of optimizing a single auxiliary function, we compute the next best solution $\mathbf{u}_{i+1}$ by exploring pseudo-bound family $\mathcal{F}_i(\mathbf{u}, \alpha)$ as discussed in the following.

Proposition 4. *Given a high-order functional $\mathcal{J}(\mathbf{u})$, a solution $\mathbf{u}_i$ at current iteration i and a pseudo-bound family $\mathcal{F}_i(\mathbf{u}, \alpha) : \{0, 1\}^{|\Omega|} \times [\alpha_{min}, \alpha_{max}] \to \mathbb{R}$, let us assume we have an optimal solution $\mathbf{u}(\alpha)$ for each parameter α:*

$$\mathbf{u}(\alpha) = \arg\min_{\mathbf{u}} \mathcal{F}_i(\mathbf{u}, \alpha). \qquad (7.29)$$

In this case, $\alpha^* = \arg\min_\alpha \mathcal{J}(\mathbf{u}(\alpha))$ yields a solution $\mathbf{u}_{i+1} := \mathbf{u}(\alpha^*)$ that guarantees that the original functional does not increase:

$$\mathcal{J}(\mathbf{u}_{i+1}) = \mathcal{J}(\mathbf{u}(\alpha^*)) \le \mathcal{J}(\mathbf{u}_i).$$

Proof. By definition of pseudo-bound, family $\mathcal{F}_i(\mathbf{u}, \alpha)$ includes at least one auxiliary function $\mathcal{F}_i(\mathbf{u}, \alpha')$, $\alpha' \in [\alpha_{\min}, \alpha_{\min}]$. Because α^* yields the best solution over the family of functions, then we have $\mathcal{J}(\mathbf{u}(\alpha^*)) \le \mathcal{J}(\mathbf{u}(\alpha'))$. Optimizing an auxiliary function guarantees the original functional does not increase: $\mathcal{J}(\mathbf{u}(\alpha')) \le \mathcal{J}(\mathbf{u}_i)$. This means:

$$\mathcal{J}(\mathbf{u}_{i+1}) = \mathcal{J}(\mathbf{u}(\alpha^*)) \le \mathcal{J}(\mathbf{u}(\alpha')) \le \mathcal{J}(\mathbf{u}_i). \qquad \square$$

The method in [25] proposed a general framework for building a pseudo-bound family for a given high-order function $\mathcal{J}(\mathbf{u})$. First, one can find at the current solution $\mathbf{u}_i$ an auxiliary function $\mathcal{A}_i(\mathbf{u})$ for $\mathcal{J}(\mathbf{u})$. Then, we add a parameter-weighted *bound relaxation*[4] term to the auxiliary function:

$$\mathcal{F}_i(\mathbf{u}, \alpha) = \mathcal{A}_i(\mathbf{u}) + \alpha \, \mathcal{B}_i(\mathbf{u}). \qquad (7.30)$$

In parametric family (7.30), $\alpha = 0$ corresponds to an auxiliary function. Therefore, given current solution $\mathbf{u}_i$, the best among all the minima of the family guarantees that the next solution $\mathbf{u}_{i+1}$ is similar or better than one that would have been obtained from optimizing auxiliary function $\mathcal{A}_i(\mathbf{u})$. While $\alpha \ne 0$ corresponds to pseudo-bounds that do not satisfy the bound condition for an auxiliary function, it may yield a better approximation of the original function, as illustrated earlier conceptually in Fig. 7.3. In practice, it is important to design a pseudo-bound family in a way that allows us to explore efficiently all the minima of the family. For instance, in the case of binary functions, choosing a submodular pseudo-bound family of the form in Eq. (7.28), with the unary term being monotone with respect to α, is practically convenient. In this case, parametric max-flow algorithms [25,40] can find all the global minima of the parametric family in low-order polynomial time. The following lists the main steps of the pseudo-bound optimization framework in [25].

---

[4] The authors of [25] called $\mathcal{B}_i(\mathbf{u})$ bound relaxation in the sense that it relaxes the upper-bound condition for an auxiliary function.

Pseudo-bound optimization:

1. Initialize segmentation $\mathbf{u}_0$;
2. Iterate the following steps until convergence:
   a. Given current solution $\mathbf{u}_i$, compute an auxiliary function $\mathcal{A}_i(\mathbf{u})$ for high-order term $\mathcal{J}(\mathbf{u})$;
   b. Augment the auxiliary function with a bound-relaxation term:

$$\mathcal{F}_i(\mathbf{u}, \alpha) = \mathcal{A}_i(\mathbf{u}) + \alpha \mathcal{B}_i(\mathbf{u});$$

   c. Find all the minima of the parametric family:

$$\mathbf{u}(\alpha) = \arg\min_{\mathbf{u}} \mathcal{F}_i(\mathbf{u}, \alpha), \text{ for } \alpha \in [\alpha_{\min}, \alpha_{\max}];$$

   d. Find the best candidate among the parametric solutions:

$$\alpha^* = \arg\min_{\alpha} \mathcal{J}(\mathbf{u}(\alpha));$$

   e. $\mathbf{u}_{i+1} \longleftarrow \mathbf{u}(\alpha^*).$

7.3.2 Examples

In the following we discuss examples of pseudo-bounds for two high-order problems: imposing a prior on the cardinality of the target segment and entropy optimization.

7.3.2.1 Cardinality-prior optimization

Priors on region cardinality (or size) can be useful in segmentation [13]. For instance, one can add the following term to standard segmentation functionals, thereby penalizing the deviation of region cardinality from a constant value c that is known *a priori:*

$$\mathcal{J}(S) = \left(c - \sum_{p \in S} 1 \right)^2 = (c - \mathbf{u}^t \mathbf{1})^2. \tag{7.31}$$

Of course, this term is not to be optimized alone. There are many solutions that correspond to the global optimum of the term in (7.31). In fact, any subset of image pixels that has cardinality c is a globally optimal solution. Many of these global solutions can be found trivially by just counting random pixels and stopping when we reach c. For a meaningful use, this cardinality prior has to be combined with other terms such model fitting

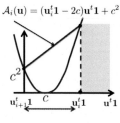

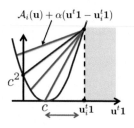

(a) (b) Bound optimization (c) Pseudo-bound optimization

Figure 7.4 Bound and pseudo-bound optimization for the cardinality penalty. (a) illustration of nested-solution constraint $(S \subset S_i)$; (b) and (c) bound and pseudo-bounds for the cardinality penalty in Eq. (7.31).

for image features and MRF regularization [13]. Here, we use this term to illustrate a simple example of pseudo-bound. Fig. 7.4 shows that the following expression is a pseudo-bound family for cardinality prior (7.31) and for nested-solution constraint $S \subset S_i$ (or equivalently $\mathbf{u}^t \mathbf{1} \le \mathbf{u}_i^t \mathbf{1}$):

$$\mathcal{F}_i(\mathbf{u}, \alpha) = \underbrace{(\mathbf{u}^t \mathbf{1} - 2c)\mathbf{u}^t \mathbf{1} + c^2}_{\mathcal{A}_i(\mathbf{u})} + \alpha \underbrace{(\mathbf{u}^t \mathbf{1} - \mathbf{u}_i^t \mathbf{1})}_{\mathcal{B}_i(\mathbf{u})}. \qquad (7.32)$$

In the next section, we will give details as to how to derive analytically general bounds for a large class of high-order functions, including this cardinality prior. For now, this simple example and its illustration in Fig. 7.4 can give us some intuition as to how pseudo-bounds might help in finding better optima than a single auxiliary function. Clearly, the linear function depicted in red in Fig. 7.4 (b), which corresponds to $\mathcal{A}_i(\mathbf{u})$ in Eq. (7.32), is an auxiliary function for the cardinality prior. For $S \subset S_i$, it is an upper bound, which is also tight at current solution $\mathbf{u}_i$. Therefore, decreasing this linear function, subject to nested-solution constraint $S \subset S_i$, guarantees that cardinality prior $\mathcal{J}(S)$ does not increase at the current iteration. The pseudo-bounds, which are depicted in green in Fig. 7.4 (c), are also linear functions but with different slopes. Exploring all these slopes, if doable computationally, and choosing the best among them, might yield a lower value and faster decrease of $\mathcal{J}$ at the current iteration.

7.3.2.2 Entropy optimization

Earlier in this chapter, we saw that the popular GrabCut algorithm can be viewed as a bound optimizer of the high-order entropy function in (7.16). The auxiliary function in this case, given in Eq. (7.17), is the sum of unary

Table 7.1 Bound (i.e., GrabCut) [33] versus pseudo-bounds [25] for optimizing the regularized entropy.

	Mean $\mathcal{E}$	Nb of lower $\mathcal{E}$	Mean time (s)	Error rate
Bound (16^3 bins)	1.2349×10^6	1	1.0	7.10%
Pseudo-bound (16^3 bins)	1.2335×10^6	38	11.2	5.80%
Bound (32^3 bins)	1.7064×10^6	2	0.9	8.78%
Pseudo-bound (32^3 bins)	1.7029×10^6	37	11.7	5.60%
Bound (64^3 bins)	2.2408×10^6	1	0.9	9.31%
Pseudo-bound (64^3 bins)	2.2361×10^6	47	14.1	5.56%

and pairwise submodular potentials. Therefore, it can be optimized with a graph cut. To examine the usefulness of pseudo-bounds experimentally in the case of two-region segmentation, the authors of [25] added to auxiliary function (7.17) a bound-relaxation term of the form $B_i(\mathbf{u}) = \alpha(\mathbf{u}^t\mathbf{1} - \mathbf{u}_i^t\mathbf{1})$. They solved the ensuing family of functions with parametric max-flow and reported experimental evaluations on the GrabCut data set [33], which contains 50 images with ground-truth segmentations. Table 7.1 summarizes the results of [25]. In these experiments, normalized histograms were used as probability models, and the results were reported for various values of the number of histogram bins. To evaluate the optimization performance, the authors of [25] used the average of function (7.16) at convergence. They also reported the number of times in which pseudo-bound optimization yielded a lower function at convergence than bound optimization.[5] These results show that pseudo-bound optimization attains consistently lower values of the regularized entropy at convergence. This improvement in optimization quality over bound optimization (or GrabCut in this case) translates into better segmentation accuracies, i.e., lower mean errors[6] over the data set.

Fig. 7.5 depicts an example of how pseudo-bound optimization increases robustness to the initial conditions and might help in avoiding weak local minima. The example uses two different initializations for the

---

[5] Obtained functions are considered equal if the absolute value of their difference is within a certain threshold.

[6] For each image, the error is the percentage of misclassified pixels with respect to the ground-truth.

Initializations	Bound optimization	Pseudo-bound optimization

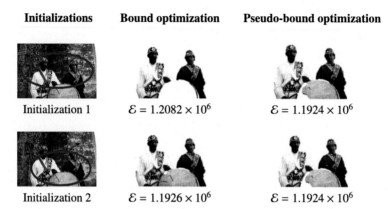

| Initialization 1 | $\mathcal{E} = 1.2082 \times 10^6$ | $\mathcal{E} = 1.1924 \times 10^6$ |
| Initialization 2 | $\mathcal{E} = 1.1926 \times 10^6$ | $\mathcal{E} = 1.1924 \times 10^6$ |

Figure 7.5 The results of bound and pseudo-bound optimization using two different initializations. This example shows that pseudo-bounds increase robustness to the initial conditions and might help to avoid weak local minima.

same image and computes the corresponding solutions. Pseudo-bound optimization yielded the same solution for both initializations, whereas bound optimization computed two different solutions.

7.4 Auxiliary functionals

In this section, we discuss how to invoke Jensen's inequality and some convexity arguments so as to derive bounds for a general class of high-order functionals [41]. We will start with objectives that take the form of some nonlinear function of a unary term, and, then, we will examine bounds for a class of fractional terms.

7.4.1 Nonlinear function of a unary term

Let us consider high-order functional of the following general form:

$$\mathcal{J}(S) = g\left(\sum_{p \in S} v_p\right) = g\left(\sum_{p \in \Omega} v_p u_p\right) = g\left(\mathbf{u}^t \mathbf{v}\right), \qquad (7.33)$$

where, recall, $\mathbf{u} = (u_p)_{p \in \Omega} \in \{0, 1\}^{|\Omega|}$ is a binary indicator vector for region S and $\mathbf{v} = (v_p)_{p \in \Omega} \in \mathbb{R}^{|\Omega|}$ a vector containing a unary potential v_p for each pixel p. Several high-order functions that we discussed earlier in this book befit this general form. For instance, the size-prior term in (7.31), which discourages a segment S from deviating from a given (known) size c [13], can be written as $(c - \mathbf{u}^t \mathbf{1})^2$, where $\mathbf{1}$ is the $|\Omega|$-dimensional vector whose components are all equal to 1. Therefore, it befits the general form in (7.33) if we use these particular functions: $\mathbf{v} = \mathbf{1}$ and $g(\gamma) = (c - \gamma)^2$.

Histogram distances, which we discussed in Chapter 6, are other examples of high-order functions that follow the general form in Eq. (7.33). Recall that minimizing this type of functions encourages the histogram of features within a region S to match a given (learned a priori) histogram $\{h_{\mathbf{b}}, \mathbf{b} \in \mathcal{Z}\}$, where $\mathbf{b}$ denotes bins (e.g., color bins) belonging to a finite set $\mathcal{Z}$. Such functions were shown to be useful in object tracking [14] and in co-segmentation of image pairs [42–44]. Typically, they are expressed as an $\mathcal{L}_j$-distance, $j \in \mathbb{N}^*$:

$$\mathcal{J}(S) = \sum_{\mathbf{b} \in Z} \left| h_{\mathbf{b}} - \sum_{p \in S} [\mathbf{f}_p = \mathbf{b}] \right|^j, \qquad (7.34)$$

where, recall, $[.]$ is the Iverson bracket, which takes value 1 if its argument is true and 0 otherwise. $\mathbf{f}_p$ denotes image features at pixel p. The linear term $\sum_{p \in S} [\mathbf{f}_p = \mathbf{b}]$ counts, within region S, the number of pixels that belong to histogram bin $\mathbf{b}$. The histogram distance in Eq. (7.34) is the sum of expressions following the general form in (7.33). Each term in this sum can be expressed as $g^{\mathbf{b}}(\mathbf{u}^t \mathbf{v}^{\mathbf{b}})$, with $g^{\mathbf{b}}(\gamma) = |\gamma - h_{\mathbf{b}}|^j$, $\mathbf{b} \in \mathcal{Z}$, and $\mathbf{v}^{\mathbf{b}} = (v_p^{\mathbf{b}})_{p \in \Omega}$ a vector containing unary potentials: $v_p^{\mathbf{b}} = [\mathbf{f}_p = \mathbf{b}]$.

The following proposition introduces the main result in [41], which provides a general auxiliary function for high-order terms of the form (7.33), with nonnegative potentials $\mathbf{v} \geq 0$ (i.e, $v_p \geq 0 \;\; \forall p$) and nested solutions. Nested solutions satisfy the following constraint: Given current solution S_i, the set of possible solutions S at the next iteration $(i+1)$ verifies $S \subset S_i$; see Fig. 7.4a for an illustration.

Proposition 5. *Assume that g is a convex function and $\mathbf{v}$ are nonnegative potentials. In this case, given current solution S_i, we have the following auxiliary function for the general high-order term in (7.33) and for any segment S verifying $S \subset S_i$:*

$$\mathcal{A}_i(\mathbf{u}) = g(\mathbf{u}_i^t \mathbf{v}) + a_i(\mathbf{u}^t \mathbf{v} - \mathbf{u}_i^t \mathbf{v}), \qquad (7.35)$$

where constant factor a_i is given as a function of the current solution:

$$a_i = \frac{1}{\mathbf{u}_i^t \mathbf{v}} \left[g(\mathbf{u}_i^t \mathbf{v}) - g(0) \right]. \qquad (7.36)$$

Notice that the auxiliary function in Eq. (7.35) is a linear (unary) term, up to the additive and multiplicative constants. Notice also that the auxiliary function we used earlier in Eq. (7.32) for the cardinality prior is a particular case of the general auxiliary function in Eq. (7.35).

Proof. The proof of Proposition 5 is based on the following lemma.

Lemma 1. *For any convex function g, and any triplet of real numbers (γ, w, z) verifying $\gamma < w < z$, we have the following inequality:*

$$g(w) \leq g(z) + \left[g(\gamma) - g(z)\right] \frac{z - w}{z - \gamma}. \tag{7.37}$$

This can be obtained from Jensen's inequality for a convex function g. Observe that $w = \frac{\gamma(z-w)}{z-\gamma} + \frac{z(w-\gamma)}{z-\gamma}$ and that coefficients $\frac{(z-w)}{z-\gamma}$ and $\frac{(w-\gamma)}{z-\gamma}$ sum to 1. Thus, applying Jensen's inequality to g gives $g(w) \leq \frac{g(\gamma)(z-w)}{z-\gamma} + \frac{g(z)(w-\gamma)}{z-\gamma}$, which, after some manipulations, yields (7.37). For nonnegative $\mathbf{v}$ and nested solutions $S \subset S_i$ (Fig. 7.4.a), we have: $0 \leq \mathbf{u}^t\mathbf{v} \leq \mathbf{u}_i^t\mathbf{v}$. Applying inequality (7.37) to $\gamma = 0$, $w = \mathbf{u}^t\mathbf{v}$ and $z = \mathbf{u}_i^t\mathbf{v}$, and after some simplifications, we get the linear bound in (7.35). ☐

7.4.2 Nonlinear function of a fractional term

We examine here regularized high-order functionals, which take the general form $\sum_{l=0}^{L-1} \mathcal{J}(S^l) + \lambda \mathcal{R}(\mathbf{S})$, with $\mathcal{J}$ expressed as a nonlinear function of the ratio of unary terms for each region $S \in \Omega$:

$$\mathcal{J}(S) = r\left(\frac{\sum_{p \in S} v_p}{\sum_{p \in S} l_p}\right) = r\left(\frac{\sum_{p \in \Omega} u_p v_p}{\sum_{p \in \Omega} u_p l_p}\right) = r\left(\frac{\mathbf{u}^t\mathbf{v}}{\mathbf{u}^t\mathbf{1}}\right). \tag{7.38}$$

$\mathbf{v} = (v_p)_{p \in \Omega}$ and $\mathbf{1} = (l_p)_{p \in \Omega}$ are two $|\Omega|$-dimensional vectors, each containing some unary potentials, and r is a convex scalar function defined over a convex domain. Several useful high-order functions that we discussed in Chapter 6 follow this general ratio form. This includes several measures of similarity between distributions [9,13,41] as well as penalties involving shape moments [45,46] or other regional statistics [12], among others. For instance, one useful problem is to find a region S so that the distribution of features within S most closely matches a given (learned) model distribution $\{m_b, \mathbf{b} \in \mathcal{Z}\}$, where $\mathbf{b}$ denotes bins (e.g., color bins) belonging to a finite set $\mathcal{Z}$. In this case, one choice is to minimize the Kullback–Leibler (KL) divergence [47]:

$$\sum_{\mathbf{b}\in\mathcal{Z}} m_{\mathbf{b}} \ln\left(\frac{m_{\mathbf{b}}}{\frac{\sum_{p\in\Omega} u_p K_p^{\mathbf{b}}}{\sum_{p\in\Omega} u_p}+\epsilon}\right) = \underbrace{\sum_{\mathbf{b}\in\mathcal{Z}} m_{\mathbf{b}} \ln m_{\mathbf{b}}}_{Constant} - \underbrace{\sum_{\mathbf{b}\in\mathcal{Z}} m_{\mathbf{b}} \ln\left(\frac{\sum_{p\in\Omega} u_p K_p^{\mathbf{b}}}{\sum_{p\in\Omega} u_p}+\epsilon\right)}_{Variable}.$$

$$(7.39)$$

In this expression, fractional term $\frac{\sum_{p\in\Omega} u_p K_p^{\mathbf{b}}}{\sum_{p\in\Omega} u_p}$ evaluates at bin **b** the KDE of the distribution of features within region S, with $K_p^{\mathbf{b}}$ a kernel function measuring a similarity between feature point $\mathbf{f}_p$ and bin **b**, e.g., the Gaussian kernel:

$$K_p^{\mathbf{b}} = \frac{1}{(2\pi\sigma^2)^{\frac{N}{2}}} \exp^{-\frac{\|\mathbf{b}-\mathbf{f}_p\|^2}{2\sigma^2}}.$$

$$(7.40)$$

Parameter σ is the width of the kernel. One can also use the normalized histogram as density estimate, which corresponds to $K_p^{\mathbf{b}} = [\mathbf{f}_p = \mathbf{b}]$. ϵ is a small positive constant, which is added to avoid division by 0. Ignoring the constant term in Eq. (7.39), it is easy to verify that, for each $\mathbf{b} \in \mathcal{Z}$, we have a function $r = r^{\mathbf{b}}$ that follows the general form in (7.38), with $v_p = K_p^{\mathbf{b}}$, $l_p = 1$ and $r^{\mathbf{b}}(y) = -m_{\mathbf{b}} \ln(y+\epsilon)$. The following proposition, introduced in [41], gives a unary auxiliary function for high-order fractional terms of the general form in (7.38).

Proposition 6. *Assume that $\mathbf{v}$ and $\mathbf{1}$ are nonnegative and r is convex, monotonically decreasing and defined at 0. In this case, we have the following auxiliary function for the general high-order term in (7.38) and for any segment S verifying $S \subset S_i$:*

$$\mathcal{A}_i(\mathbf{u}) = r\left(\frac{\mathbf{u}_i^t \mathbf{v}}{\mathbf{u}_i^t \mathbf{1}}\right) + b_i(\mathbf{u}^t \mathbf{v} - \mathbf{u}_i^t \mathbf{v}),$$

$$(7.41)$$

where constant factor b_i is given as a function of the current solution:

$$b_i = \frac{1}{\mathbf{u}_i^t \mathbf{v}}\left[r\left(\frac{\mathbf{u}_i^t \mathbf{v}}{\mathbf{u}_i^t \mathbf{1}}\right) - r(0)\right].$$

$$(7.42)$$

Proof. Because $S \subset S^i$ and l_p is nonnegative, we have $\sum_S l_p \le \sum_{S^i} l_p$, i.e., $\mathbf{u}^t\mathbf{1} \le \mathbf{u}_i^t\mathbf{1}$. This means that, when function r is monotonically decreasing and $\mathbf{u}^t\mathbf{v}$ is nonnegative, the high-order term of the general form in (7.38)

Initialization   Bound optim. ($\mathcal{E}$ = 7265)   Pseudo-bound ($\mathcal{E}$) = 6452)

Figure 7.6 A distribution-matching example based on the KL divergence. (*Figures from [25] © Springer.*)

is upper-bounded as follows:

$$r\left(\frac{\mathbf{u}^t\mathbf{v}}{\mathbf{u}^t\mathbf{1}}\right) \le r\left(\frac{\mathbf{u}^t\mathbf{v}}{\mathbf{u}_i^t\mathbf{1}}\right) = r_i(\mathbf{u}^t\mathbf{v}), \tag{7.43}$$

where function r_i is given by:

$$r_i(\gamma) = r\left(\frac{\gamma}{\mathbf{u}_i^t\mathbf{1}}\right) \tag{7.44}$$

Notice that function r_i is convex when r is convex. Therefore, the upper bound in Eq. (7.43) is a convex function of a unary term, which enables us to apply the result in Proposition 5 to r_i. This yields the result in Proposition 6. □

Notice that, at each iteration, auxiliary function (7.41) is a unary term. Therefore, for regularized high-order terms of the form $\sum_{l=0}^{L-1} \mathcal{J}(S^l) + \lambda\mathcal{R}(\mathbf{S})$, one can replace each $\mathcal{J}(S^l)$ by the corresponding auxiliary function and solve the overall problem with graph cuts.

7.4.2.1 Example

Fig. 7.6 depicts a distribution matching example from [25]. The example is based on the KL divergence between normalized histograms, which is used in conjunction with a standard pairwise edge-sensitive MRF regularization. Optimization is carried out using the auxiliary function in Eq. (7.41), with $r = r^{\mathbf{b}}(\gamma) = -m_{\mathbf{b}}\ln(\gamma+\epsilon)$ for each term in the KL divergence. This auxiliary function is augmented with a bound-relaxation term $\mathcal{B}_i(\mathbf{u}) = \alpha(\mathbf{u}^t\mathbf{1} - \mathbf{u}_i^t\mathbf{1})$ and combined with the MRF term. It is optimized with a graph cut in the case of bound optimization ($\alpha = 0$) and with parametric max-flow in the case of pseudo-bound optimization ($\alpha \in \mathbb{R}$). The model distribution is

Table 7.2 Bound [41] versus pseudo-bound [25] for distribution matching using the KL divergence in conjunction with a pairwise MRF regularization.

Optimizer	Mean $\mathcal{E}$	Mean error	Mean time (s)
Bound	6189	16.54%	1.80 s
Pseudo-bound	5849	3.63%	2.98 s

computed from the ground-truth segmentation of the image. This is not a realistic assumption in practice because the ground-truth segmentation is not known. However, the purpose here is to evaluate the optimization performance. The example in Fig. 7.6 shows that the use of pseudo-bounds improved the value of the objective function at convergence, i.e., yielded a lower value. This translated into a segmentation result that is more consistent with the ground-truth. Table 7.2 summarizes the experimental evaluations of [25], which performed the same experiment on all the 50 images of the GrabCut data set [33], and computed the average objective function obtained at convergence. These results confirm how pseudo-bounds improve the optimization performance, without adding significantly to the computational time. This improvement in optimization quality over bound optimization also translates into better segmentation errors (on average).

References

[1] Y. Boykov, O. Veksler, R. Zabih, Fast approximate energy minimization via graph cuts, IEEE Transactions on Pattern Analysis and Machine Intelligence 23 (11) (2001) 1222–1239.

[2] Y. Boykov, V. Kolmogorov, An experimental comparison of min-cut/max-flow algorithms for energy minimization in vision, IEEE Transactions on Pattern Analysis and Machine Intelligence 26 (9) (2004) 1124–1137.

[3] Y. Boykov, M.P. Jolly, Interactive graph cuts for optimal boundary and region segmentation of objects in n-d images, in: IEEE International Conference on Computer Vision (ICCV), 2001, pp. 105–112.

[4] Y. Boykov, G. Funka Lea, Graph cuts and efficient n-d image segmentation, International Journal of Computer Vision 70 (2) (2006) 109–131.

[5] A. Blake, P. Kohli, C. Rother, Markov Random Fields for Vision and Image Processing, MIT Press, 2011.

[6] P. Krähenbühl, V. Koltun, Efficient inference in fully connected CRFs with Gaussian edge potentials, in: Advances in Neural Information Processing Systems (NIPS), 2011, pp. 109–117.

[7] M. Tang, D. Marin, I. Ben Ayed, Y. Boykov, Normalized cut meets MRF, in: European Conference on Computer Vision (ECCV), Part II, 2016, pp. 748–765.

[8] T. Taniai, Y. Matsushita, T. Naemura, Superdifferential cuts for binary energies, in: IEEE Conference on Computer Vision and Pattern Recognition (CVPR), 2015, pp. 2030–2038.

[9] I. Ben Ayed, K. Punithakumar, S. Li, Distribution matching with the Bhattacharyya similarity: A bound optimization framework, IEEE Transactions on Pattern Analysis and Machine Intelligence 37 (9) (2015) 1777–1791.

[10] Y. Boykov, H.N. Isack, C. Olsson, I. Ben Ayed, Volumetric bias in segmentation and reconstruction: Secrets and solutions, in: IEEE International Conference on Computer Vision (ICCV), 2015, pp. 1769–1777.

[11] Y. Kee, M. Souiai, D. Cremers, J. Kim, Sequential convex relaxation for mutual information-based unsupervised figure-ground segmentation, in: IEEE Conference on Computer Vision and Pattern Recognition (CVPR), 2014, pp. 4082–4089.

[12] Y. Lim, K. Jung, P. Kohli, Efficient energy minimization for enforcing label statistics, IEEE Transactions on Pattern Analysis and Machine Intelligence 36 (9) (2014) 1893–1899.

[13] L. Gorelick, F.R. Schmidt, Y. Boykov, Fast trust region for segmentation, in: IEEE Conference on Computer Vision and Pattern Recognition (CVPR), 2013, pp. 1714–1721.

[14] H. Jiang, Linear solution to scale invariant global figure ground separation, in: IEEE Conference on Computer Vision and Pattern Recognition (CVPR), 2012, pp. 678–685.

[15] V.Q. Pham, K. Takahashi, T. Naemura, Foreground-background segmentation using iterated distribution matching, in: IEEE International Conference on Computer Vision and Pattern Recognition (CVPR), 2011, pp. 2113–2120.

[16] J.H. Kappes, B. Andres, F.A. Hamprecht, C. Schnörr, S. Nowozin, D. Batra, et al., A comparative study of modern inference techniques for structured discrete energy minimization problems, International Journal of Computer Vision 115 (2) (2015) 155–184.

[17] L. Gorelick, Y. Boykov, O. Veksler, I. Ben Ayed, A. Delong, Local submodularization for binary pairwise energies, IEEE Transactions on Pattern Analysis and Machine Intelligence 39 (10) (2017) 1985–1999.

[18] V. Kolmogorov, T. Schoenemann, Generalized sequential tree-reweighted message pass, arXiv:1205.6352, 2012, 17 pp.

[19] J. Shi, J. Malik, Normalized cuts and image segmentation, IEEE Transactions on Pattern Analysis and Machine Intelligence 22 (8) (2000) 888–905.

[20] U. Von Luxburg, A tutorial on spectral clustering, Statistics and Computing 17 (4) (2007) 395–416.

[21] J. Dolz, I. Ben Ayed, C. Desrosiers, Unbiased shape compactness for segmentation, in: Medical Image Computing and Computer Assisted Intervention (MICCAI), Part I, 2017, pp. 755–763.

[22] M. Niethammer, C. Zach, Segmentation with area constraints, Medical Image Analysis 17 (1) (2013) 101–112.

[23] H. Kervadec, J. Dolz, M. Tang, E. Granger, Y. Boykov, I. Ben Ayed, Constrained-CNN losses for weakly supervised segmentation, Medical Image Analysis 54 (2019) 88–99.

[24] C. Nieuwenhuis, E. Töppe, L. Gorelick, O. Veksler, Y. Boykov, Efficient squared curvature, in: IEEE Conference on Computer Vision and Pattern Recognition (CVPR), 2014, pp. 4098–4105.

[25] M. Tang, I. Ben Ayed, Y. Boykov, Pseudo-bound optimization for binary energies, in: European Conference on Computer Vision (ECCV), Part V, 2014, pp. 691–707.

[26] Z. Zhang, J.T. Kwok, D.Y. Yeung, Surrogate maximization/minimization algorithms and extensions, Machine Learning 69 (2007) 1–33.

[27] M. Fashing, C. Tomasi, Mean shift is a bound optimization, IEEE Transactions on Pattern Analysis and Machine Intelligence 27 (2005) 471–474.

[28] A.L. Yuille, A. Rangarajan, The concave-convex procedure (CCCP), in: Advances in Neural Information Processing Systems (NIPS), 2001, pp. 1033–1040.

[29] M. Narasimhan, J.A. Bilmes, A submodular-supermodular procedure with applications to discriminative structure learning, in: Conference on Uncertainty in Artificial Intelligence (UAI), 2005, pp. 404–412.

[30] K. Lange, D.R. Hunter, I. Yang, Optimization transfer using surrogate objective functions, Journal of Computational and Graphical Statistics 9 (1) (2000) 1–20.

[31] D.D. Lee, H.S. Seung, Algorithms for non-negative matrix factorization, in: NIPS, 2000, pp. 556–562.

[32] L. Gorelick, Y. Boykov, O. Veksler, Adaptive and move making auxiliary cuts for binary pairwise energies, in: IEEE Conference on Computer Vision and Pattern Recognition (CVPR), 2017, pp. 6062–6070.

[33] C. Rother, V. Kolmogorov, A. Blake, GrabCut: Interactive foreground extraction using iterated graph cuts, ACM Transactions on Graphics 23 (3) (2004) 309–314.

[34] T. Chan, L. Vese, Active contours without edges, IEEE Transactions on Image Processing 10 (2) (2001) 266–277.

[35] M. Tang, D. Marin, I. Ben Ayed, D. Marin, Y. Boykov, Kernel cuts: Kernel & spectral clustering meet regularization, International Journal of Computer Vision 127 (5) (2019) 477–511.

[36] D. Marin, M. Tang, I. Ben Ayed, Y. Boykov, Kernel clustering: Density biases and solutions, IEEE Transactions on Pattern Analysis and Machine Intelligence 41 (1) (2018) 136–147.

[37] M. Tang, I. Ben Ayed, D. Marin, Y. Boykov, Secrets of GrabCut and kernel K-means, in: IEEE International Conference on Computer Vision (ICCV), 2015, pp. 1555–1563.

[38] M. Tang, L. Gorelick, O. Veksler, Y. Boykov, GrabCut in one cut, in: International Conference on Computer Vision (ICCV), 2013, pp. 1769–1776.

[39] S.E. Chew, N.D. Cahill, Semi-supervised normalized cuts for image segmentation, in: IEEE International Conference on Computer Vision (ICCV), 2015, pp. 1716–1723.

[40] V. Kolmogorov, Y. Boykov, C. Rother, Applications of parametric maxflow in computer vision, in: IEEE International Conference on Computer Vision (ICCV), 2007, pp. 1–8.

[41] I. Ben Ayed, L. Gorelick, Y. Boykov, Auxiliary cuts for general classes of higher-order functionals, in: IEEE International Conference on Computer Vision and Pattern Recognition (CVPR), 2013, pp. 1304–1311.

[42] C. Rother, V. Kolmogorov, T. Minka, A. Blake, Cosegmentation of image pairs by histogram matching – incorporating a global constraint into MRFs, in: IEEE International Conference on Computer Vision and Pattern Recognition (CVPR), 2006, pp. 993–1000.

[43] L. Mukherjee, V. Singh, C.R. Dyer, Half-integrality based algorithms for cosegmentation of images, in: IEEE Conference on Computer Vision and Pattern Recognition (CVPR), 2009, pp. 2028–2035.

[44] S. Vicente, V. Kolmogorov, C. Rother, Cosegmentation revisited: Models and optimization, in: European Conference on Computer Vision (ECCV), Part 2, 2010, pp. 465–479.

[45] M. Klodt, D. Cremers, A convex framework for image segmentation with moment constraints, in: IEEE International Conference on Computer Vision (ICCV), 2011, pp. 2236–2243.

[46] A. Foulonneau, P. Charbonnier, F. Heitz, Multi-reference shape priors for active contours, International Journal of Computer Vision 81 (1) (2009) 68–81.

[47] A. Mitiche, I. Ben Ayed, Variational and Level Set Methods in Image Segmentation, 1st ed., Springer, 2011, 192 pp.

CHAPTER 8

Trust-region optimization

8.1 General-form problem

As discussed in great details in the previous chapters, high-order functions are useful in a wide range of computer vision, image processing and data analysis tasks. Recall that, for binary segmentation,[1] a high-order term could be expressed in the following general form:

$$\mathcal{J}(\mathbf{u}) = g\left(\mathbf{u}^t \mathbf{v}^1, \ldots, \mathbf{u}^t \mathbf{v}^J\right), \tag{8.1}$$

where g is a nonlinear function and:

- $\mathbf{u} = (u_p)_{p \in \Omega} \in \{0, 1\}^{|\Omega|}$ is a binary vector that contains the indicator variables of the foreground region $S \subset \Omega$ (Ω is the spatial image domain): $u_p = 1$ if pixel p belongs to foreground S and $u_p = 0$ otherwise (i.e., p belongs to background $\Omega \setminus S$).
- $\forall j \in 1, \ldots, J$, $\mathbf{v}^j = (v_p^j)_{p \in \Omega} \in \mathbb{R}^{|\Omega|}$ is a vector containing a unary potential v_p^j for each pixel p. Each dot product $\mathbf{u}^t \mathbf{v}^j$ is a unary term, which could be expressed as a summation over the foreground region as follows:

$$\mathbf{u}^t \mathbf{v}^j = \sum_{p \in S} v_p^j = \sum_{p \in \Omega} v_p^j u_p.$$

Several high-order functions that we discussed in detail earlier in this book befit this general form, i.e., a nonlinear combination of unary-potential terms $\mathbf{u}^t \mathbf{v}^j$, $j = 1, \ldots, J$. We examined several examples of high-order terms $\mathcal{J}(\mathbf{u})$ of the general form in Eq. (8.1) and illustrated the usefulness of such terms in practice. This included, for instance, priors on the distributions of features within the target regions [2], or priors on the geometric shape moments [1,3] and sizes [4,5] of the target regions. Such priors on region shape/size could very useful in medical-imaging applications, for instance. Unfortunately, powerful and global optimization techniques, such as graph cuts [6,7], are restricted to special forms of regional and boundary functions, e.g., g in Eq. (8.1) is a linear (unary) or quadratic (pairwise) function

---

[1] In this chapter, to facilitate the presentation, we focus our discussion on the two-region segmentation case.

High-Order Models in Semantic Image Segmentation
https://doi.org/10.1016/B978-0-12-805320-1.00013-0

satisfying submodularity conditions. They are not applicable to the more general case of (8.1), where g is an arbitrary nonlinear function and objective $\mathcal{J}$ is a high-order, nonsubmodular function. In the following, we consider the general case, where an overall objective $\mathcal{E}$ integrates a high-order term $\mathcal{J}$ of the form (8.1) with a pairwise submodular function $\mathcal{F}$ amenable to global graph-cut optimization:

$$\mathcal{E}(\mathbf{u}) = \mathcal{J}(\mathbf{u}) + \mathcal{F}(\mathbf{u}). \tag{8.2}$$

Later in this chapter, we will examine a specific example, where high-order function $\mathcal{J}(\mathbf{u})$ encodes a shape-compactness prior and $\mathcal{F}$ is the sum of standard unary and pairwise MRF regularization potentials (as discussed in great detail earlier in the book).

8.2 Trust-region optimization

Trust-region algorithms [1,8,9] belong to a general class of iterative optimization procedures that splits a difficult problem into a sequence of easier substeps. Within each iteration, an approximation of the objective is built in the vicinity of the current solution. In general, the approximation fits the initial problem only locally. Therefore, it is "trusted" only within a region in the vicinity of the current solution, often referred to as "trust region". Of course, the approximation should be easier to optimize than the initial objective (e.g., a linear, first-order Taylor approximation). As we will see later with a specific example, one could derive unary (linear) approximations of general high-order functions of the form (8.1) using first-order Gateâux derivatives. Hence, at each iteration, one could optimize globally the approximation to obtain candidate solutions constrained to be within the trust region, an iterative substep often referred to as trust region subproblem. At each iteration, the obtained candidate solution is evaluated and the size of the trust region is updated, using some measure of the quality of the current approximation. There are several variants of trust-region methods. The differences between such variants could be in the type of the approximation, the optimizer used to solve the trust-region subproblems and/or the criterion used for updating the trust-region size and rejecting/accepting candidate solutions. In the following, we will discuss the trust-region approach introduced in [1], which was devised specifically for image segmentation. A more detailed discussion of the general trust-region framework can be found in [8].

Assume that we want to minimize a function $\mathcal{E}(\mathbf{u})$ taking the general form in (8.2). At each iteration i, and given current solution $\mathbf{u}_i$, we approximate objective $\mathcal{E}(\mathbf{u})$ as follows:

$$\mathcal{E}_i(\mathbf{u}) = \mathcal{J}_i(\mathbf{u}) + \mathcal{F}(\mathbf{u}), \tag{8.3}$$

where $\mathbf{u} \in [0, 1]^{|\Omega|}$ is soft (not binary) segmentation, i.e., integer constraints $u_p \in \{0, 1\}$ are relaxed to $u_p \in [0, 1]$, and $\mathcal{J}_i(\mathbf{u})$ is the first-order (unary) Gateâux derivative approximation of high-order term $\mathcal{J}(\mathbf{u})$ at the vicinity of current solution $\mathbf{u}_i$. Let distance d_i denotes the size of the trust region, which is adjusted automatically at the previous iteration (more details on this will be provided in what follows). Then, at each iteration i, we can solve the following constrained subproblem:

$$\min_{\mathbf{u} \in [0,1]^{|\Omega|}} \mathcal{E}_i(\mathbf{u}) \quad \text{s.t.} \quad \mathcal{D}(\mathbf{u}, \mathbf{u}_i) < d_i, \quad \text{for } i = 0, 1, 2, \ldots, \tag{8.4}$$

where $\mathcal{D}(\mathbf{u}, \mathbf{u}_i)$ is some measure evaluating the discrepancy between the segmentation defined by variable $\mathbf{u}$ and current solution $\mathbf{u}_i$. Of course, different choices of $\mathcal{D}$ and/or different ways for solving the constrained subproblems in (8.4) lead to different variants of the trust-region approach. Later in this chapter, we will discuss specific choices that were introduced in [1] in the context of image segmentation. Let us first summarize the general trust-region framework in the following.

8.2.1 General trust-region framework

The general trust-region framework is based on the following steps:
1. Initialize segmentation variable ($\mathbf{u}_0$);
2. Initialize trust-region size (d_0);
3. Iterate until convergence:
   a. Solve the trust region subproblem:

$$\mathbf{u}^* \longleftarrow \underset{\mathcal{D}(\mathbf{u},\mathbf{u}_i)<d_i}{\text{argmin}} \, \mathcal{E}_i(\mathbf{u})$$

   b. Approximate decrease of the functional: $\widetilde{\Delta \mathcal{E}} = \mathcal{E}_i(\mathbf{u}_i) - \mathcal{E}_i(\mathbf{u}^*)$;
   c. Actual decrease of the functional: $\Delta \mathcal{E} = \mathcal{E}(\mathbf{u}_i) - \mathcal{E}(\mathbf{u}^*)$;
   d. Update the current solution:

$$\mathbf{u}_{i+1} \longleftarrow \begin{cases} \mathbf{u}^* & \text{if } \frac{\Delta \mathcal{E}}{\widetilde{\Delta \mathcal{E}}} > \tau_1 \\ \mathbf{u}_i & \text{otherwise;} \end{cases}$$

e. Adjust the trust region:

$$d_{i+1} \longleftarrow \begin{cases} d_i \gamma & \text{if } \frac{\Delta \mathcal{E}}{\Delta \tilde{\mathcal{E}}} > \tau_2 \\ d_i / \gamma & \text{otherwise.} \end{cases}$$

Assume that we have an efficient solver for the trust-region subproblem in 3.a (more details on this will follow), which yields a candidate solution $\mathbf{u}^*$. Given the latter, the quality of the approximation is evaluated via the ratio between the actual and approximate decrease of the functional. Depending on the value of this ratio, we update the solution (step 3.d) as well as the trust region (step 3.e). Parameters τ_1, τ_2 and γ are set experimentally. Commonly, parameter τ_1 in step 3.d is set equal to zero. In this case, candidate solutions that improve the actual functional are accepted. For parameter τ_2 (step 3.e), a standard choice in trust-region methods is $\tau_2 = 0.25$ [9]. This means that a decrease ratio higher than this value is assumed to indicate a good approximation, thereby allowing a larger trust-region size (i.e., a larger step size), via multiplying distance d_i by a factor $\gamma > 1$.

8.2.2 Lagrangian formulation of the subproblems

The main part of the trust-region framework just described is solving the constrained-optimization problem in step 3.a. One could solve each of these subproblems via an unconstrained Lagrangian formulation, which minimizes:

$$\mathcal{L}_i = \mathcal{E}_i(\mathbf{u}) + \frac{\lambda}{2} \mathcal{D}(\mathbf{u}, \mathbf{u}_i). \tag{8.5}$$

Of course, the choice of distance $\mathcal{D}$ might affect the difficulty of solving problem (8.5). In the context of segmentation, the authors of [1] proposed to use a specific nonsymmetric distance defined over the space of shapes (i.e., over segmentation boundaries), which can be approximated with a unary term. Therefore, subproblem (8.5) could be solved globally with a graph cut in low-order polynomial time [7], assuming pairwise function $\mathcal{F}$ is submodular.

Let $S \subset \Omega$ denotes the segmentation region associated with binary variable $\mathbf{u}$, and $S_i \subset \Omega$ the region associated with current solution $\mathbf{u}_i$. Consider the following nonsymmetric measure over the space of shapes, which evaluates the discrepancy between two segmentation boundaries[2] ∂S and ∂S_i

---

[2] For a given segmentation region $S \in \Omega$, boundary ∂S could be defined as the set of pixels $p \in S$ such that p has one or more neighbors that do not belong to S.

[10]:

$$\text{dist}(\partial S_i, \partial S) = \sqrt{\sum_{p \in \partial S_i} \|\gamma_{\partial S}(p) - p\|^2}, \tag{8.6}$$

where, for each point p that belongs to boundary ∂S_i, $\gamma_{\partial S}(p)$ denotes the corresponding point on boundary ∂S, along the direction normal to ∂S_i, i.e., $\gamma_{\partial S}(p)$ is the intersection of ∂S and the line that is normal to ∂S_i at p. The authors of [10] showed that nonsymmetric shape discrepancy (8.6) could be approximated with a summation of unary potentials over segmentation region S, up to an additive constant, using a signed-distance representation of boundary ∂S_i:

$$\text{dist}(\partial S_i, \partial S)^2 \approx 2\sum_{p \in S} \phi_i(p) - 2\sum_{p \in S_i} \phi_i(p) \stackrel{c}{=} 2\sum_{p \in \Omega} \phi_i(p) u_p, \tag{8.7}$$

where symbol $\stackrel{c}{=}$ means up to an additive constant independent of segmentation variable $\mathbf{u}$, and ϕ_i is the signed-distance function corresponding to boundary ∂S_i:

$\phi_i(p) = -\|p - z_{\partial S_i}(p)\|$ if $p \in S_i$ and $\phi_i(p) = \|p - z_{\partial S_i}(p)\|$ otherwise.

In this distance-function representation of boundary ∂S_i, $\|p - q\|$ denotes the spatial Euclidean distance between two pixels p and q, and $z_{\partial S_i}(p)$ denotes the nearest point to p among all the points on contour ∂S_i. Clearly, (8.7) is a unary term. Therefore, when we set $\mathcal{D}(\mathbf{u}, \mathbf{u}_i) = \text{dist}(\partial S_i, \partial S)^2$, with approximation $\mathcal{E}_i$ verifying the submodularity condition, the unconstrained Lagrangian formulation in (8.5) could be solved globally with a graph cut.

The general trust-region framework detailed in the previous section adjusts the size of the trust region via inequality-constraint parameter d_i; see step 3.e in the general framework just discussed. However, with the unconstrained formulation in (8.5), we do not have direct access to parameter d_i. Instead of d_i, in (8.5), we now have multiplier λ. Therefore, in step 3.e in the general discussed trust-region framework, we cannot update d_i. Nonetheless, there is a direct correspondence between multiplier λ and d_i. For a given λ, let $\mathbf{u}_\lambda$ denotes the solution of unconstrained minimization problem (8.5). As discussed in [1], one could show that, for each λ, there is a corresponding d_i such that $\mathbf{u}_\lambda$ is also the solution of (8.4). In fact, it is possible to establish an analytical relationship between λ and d_i and show a proportionality between λ and the corresponding $1/d_i$ [1]. This is intuitive:

The smaller d_i in the inequality constraint in (8.4), the more weight (λ) we give to the distance penalty in the unconstrained Lagrangian formulation in (8.5), thereby encouraging small distances. Therefore, instead of multiplying d_i by a factor γ in step 3.e in the general trust-region framework, one could simply divide λ by the same factor. This amounts to replacing step 3.e by the following:

$$\lambda \longleftarrow \begin{cases} \lambda/\gamma & \text{if } \frac{\Delta\mathcal{E}}{\Delta\hat{\mathcal{E}}} > \tau_2 \\ \lambda\gamma & \text{otherwise.} \end{cases}$$

8.3 A shape prior example

Embedding shape priors in medical image segmentation is necessary in numerous applications. This is often the case when the target segmentation regions have intensity profiles that are very similar to other parts in the image; see the example in Fig. 8.1. In the following, we discuss a specific example of a functional of the general form in (8.2), with $\mathcal{J}$ a high-order term enforcing a shape-compactness prior on the solution and $\mathcal{F}$ containing standard unary and pairwise MRF regularization potentials (as discussed in detail in Chapter 1). Shape prior $\mathcal{J}$ is expressed as follows [11]:

$$\mathcal{J}(\mathbf{u}) = \frac{\sum_{p\in\Omega} \|p - c\|^2 u_p}{\left(\sum_{p\in\Omega} u_p\right)^2}, \tag{8.8}$$

where $\|p - c\|$ evaluates the spatial Euclidean distance between pixel p and a reference point c. Reference point c could be either a variable that depends on the segmentation (i.e., c is dependent on $\mathbf{u}$) or is fixed. In the latter case, c could be, for instance, provided by the user as input. The following particular choice of reference point c is of special interest in practice:

$$c = \frac{\sum_{p\in\Omega} p u_p}{\sum_{p\in\Omega} u_p}. \tag{8.9}$$

This expression of c corresponds the centroid of the segmentation region defined by $\{p|u_p = 1\}$. In fact, with this particular choice of c, $\mathcal{J}(\mathbf{u})$ becomes one of the well-known Hu's shape moments, which are invariant with respect to translation, rotation and scaling. Furthermore, in this case, invariant measure $\mathcal{J}(\mathbf{u})$ could be used to assess the circularity of a shape [12]. It evaluates the deviation of a given shape from the most compact shape (i.e., a circle): the lower the value of $\mathcal{J}(\mathbf{u})$, the closer the shape of the segmentation defined by $\mathbf{u}$ to a circle. Of course, this high-order shape-compactness term could not be used alone. In the experimental example we discuss next, we use $\mathcal{J}$ in conjunction with a submodular function $\mathcal{F}$, which contains standard intensity (unary) and boundary-smoothness (pairwise) terms [13]:

$$\mathcal{F}(\mathbf{u}) = \alpha \mathcal{L}(\mathbf{u}) + \beta \mathcal{R}(\mathbf{u}), \tag{8.10}$$

where $\mathcal{L}(\mathbf{u})$ is unary term, which evaluates pixel log-likelihoods, given fixed (learned *a priori*) model distributions of intensity within the target segment (foreground) and its complement (background): $\mathcal{L}(\mathbf{u}) = \sum_{p \in \Omega} u_p \left[\Pr(f_p|0) - \log \Pr(f_p|1) \right]$, with f_p denoting the observed image feature (e.g., intensity) for pixel p in Ω. $\mathcal{R}(\mathbf{u})$ is a submodular pairwise regularization term, which encourages smooth, edge-aligned segmentation boundaries: $\mathcal{R}(\mathbf{u}) = \sum_{\{p,q\} \in \mathcal{N}} \frac{1}{\|p-q\|} \exp(-\rho \|f_p - f_q\|^2) [u_p \neq u_q]$, where $[u_p \neq u_q]$ denotes Iverson bracket, taking value 1 if its argument is true and 0 otherwise. $\mathcal{N}$ is a neighborhood system containing all pairs $\{p, q\}$ of neighboring pixels (e.g., 4-, 8-, or 16-neighborhood system), and α, β and ρ are positive constants that have to be fixed empirically.

8.3.1 Gateâux derivative approximation

The shape-compactness prior previously discussed is a high-order *fractional* functional, which is not directly amenable to standard combinatorial optimization techniques. One could adopt the iterative trust-region paradigm discussed earlier in this chapter and, at each iteration, use a first-order *Gateâux* derivative approximation of $\mathcal{J}$ in the vicinity of current solution $\mathbf{u}_i$ (which we denoted $\mathcal{J}_i$ earlier). This gives an approximation of the overall objective of the form (8.3) and a trust-region subproblem of the form (8.4), which could be solved via a Lagrangian formulation and a graph cut. This is due to the fact that *Gateâux* derivative approximation $\mathcal{J}_i$ takes the form of a unary term. Hence, the approximation of the overall objective correspond to a sum of unary and pairwise submodular potentials. In Section 8.4, we will provide details for deriving first-order approximations of high-order functions of the general form in (8.1). Also, as an example, we will provide a detailed derivation for the shape-compactness term in (8.8). For the latter, expression of the first-order approximation at the vicinity of current solution $\mathbf{u}_i$ is given in the ensuing Eq. (8.20), in Section 8.4.

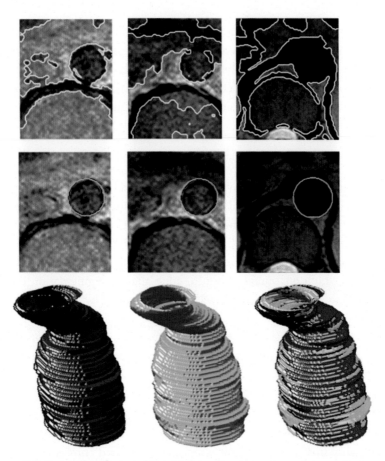

Figure 8.1 Example of use of the high-order shape term for semiautomatic segmentation of the abdominal aorta structure in magnetic resonance imaging.

8.3.2 An experimental example

Fig. 8.1 depicts a example of use of the high–order shape compactness term previously discussed. Here, the task is semiautomatic segmentation of the abdominal aorta structure in magnetic resonance imaging.[3] The problem is difficult due to the similarities in intensity between the target region (the abdominal aorta in this case) and the surrounding regions; see the example in Fig. 8.1. Also, the image edges (or contrasts) corresponding to

---

[3] Segmentation of the abdominal aorta structure in medical imaging could be useful in assessing abdominal aortic aneurysms.

the aorta boundary are weak. Furthermore, this specific tubular structure may have sudden and unpredictable changes in the size of the 2D aorta cross sections due to abnormalities such aneurysms. The first two rows in Fig. 8.1 illustrate the effect of adding the shape-compactness term in (8.8) to the standard objective in (8.10), which contains intensity (unary) and boundary-smoothness (pairwise) terms [13]. The first row depicts cross-sectional segmentation boundaries for one subject at different 2D slices, which were obtained without the high-order shape term. This corresponds to the popular model in [13]. In this case, the obtained aorta region included erroneously several parts of the background. Adding the shape term corrected the results, yielding segmentations that are consistent with the ground truth, as illustrated by the examples in the second row of Fig. 8.1. In these examples, the red curves correspond to the ground-truth segmentation, and the green curves to the segmentation results obtained by adding the shape prior. The last row of the figure shows a 3D illustration of the result obtained for the whole 3D magnetic resonance imaging volume of one subject, including the ground-truth 3D surface in red and the obtained surface in green. This 3D example illustrates how the model can deal with abrupt and unpredictable changes in the sizes of the 2D aorta cross sections, due to the scale-invariance property of the shape compactness term. In this example, the centroid of the aorta cross section within each 2D slice, i.e., c in the shape term in (8.8), is assumed to be provided by the user.[4] Of course, one could also assume c is a variable to be estimated automatically along with segmentation variable $\mathbf{u}$, via expression (8.9), thereby removing the assumption that c is given and further automating the process. This makes the high-order shape term more complex. However, the general trust-region framework discussed in this chapter is still applicable. In this experimental example, the fixed intensity models required for unary term $\mathcal{L}(\mathbf{u})$ in (8.10) are learned from inside/outside a small disc centered at reference point c, with a radius equal to 10 pixels.

8.4 Details of the Gateâux derivatives

In this section, we provide details of the *Gateâux* derivative approximations of high-order functions. We give expressions for high-order functions of

---

[4] For this type of 3D tubular/vascular structure segmentation problems, it is common to assume that an approximate centerline of the structure is provided as user input, which provides an approximate region centroid for each cross-sectional 2D slice.

the general form in (8.1) and, as example, we further provide a detailed derivation for the shape-compactness term in (8.8). Let us first consider some basic expressions and notations, which will be repeatedly used in the following. As the following details involve functional derivatives, we assume that the segmentation variable is a continuous functions $u : \Omega \subset \mathbb{R}^2 \to [0, 1]$. Also, in the following, finite summations of unary potentials v_p over the target region, i.e., $\sum_{p\in\Omega} v_p u_p = \sum_{p\in S} v_p$, are replaced with integrals of scalar functions $v : \Omega \to \mathbb{R}$:

$$\langle v, u \rangle = \int_{\Omega} v(p)u(p)dp. \tag{8.11}$$

The Gateâux derivative of the basic functional in (8.11) is:

$$\frac{\partial \langle v, u \rangle}{\partial u} = v. \tag{8.12}$$

For a functional of the following general form:

$$\mathcal{J}(u) = g\left(\langle v^1, u \rangle, \ldots, \langle v^J, u \rangle\right), \tag{8.13}$$

where $v^j : \Omega \to \mathbb{R}$ is a scalar function, $j \in 1, \ldots, J$, we have the following Gateâux derivative:

$$\frac{\partial \mathcal{J}}{\partial u} = \sum_{j=1}^{J} \frac{\partial \mathcal{J}}{\partial \langle v^j, u \rangle} \frac{\partial \langle v^j, u \rangle}{\partial u} = \sum_{j=1}^{J} \frac{\partial \mathcal{J}}{\partial \langle v^j, u \rangle} v^j. \tag{8.14}$$

Note that functional (8.13) can be viewed as a continuous version of high-order objective (8.1). Now, using derivative (8.14), the first-order approximation of a functional of the general form in (8.13) at the vicinity of current (fixed) solution u_i can be written as follows:

$$\mathcal{J}_i(u) = \mathcal{J}(u_i) + \sum_{j=1}^{J} \frac{\partial \mathcal{J}}{\partial \langle v^j, u \rangle}(u_i)\langle v^j, u - u_i \rangle \tag{8.15}$$

$$= \underbrace{\mathcal{J}(u_i) - \sum_{j=1}^{J} \frac{\partial \mathcal{J}}{\partial \langle v^j, u \rangle}(u_i)\langle v^j, u_i \rangle}_{\text{constant w.r.t. } u} + \underbrace{\sum_{j=1}^{J} \frac{\partial \mathcal{J}}{\partial \langle v^j, u \rangle}(u_i)\langle v^j, u \rangle}_{\text{unary potential}}.$$

Now, replacing integrals of the form $\langle v^j, u \rangle$ in first-order approximation (8.15) by finite summations $\mathbf{u}^t \mathbf{v}^j = \sum_{p\in\Omega} v_p^j u_p$, we obtain the following unary approximation of high-order objective (8.1) at the vicinity of current solution $\mathbf{u}_i$:

$$\mathcal{J}_i(\mathbf{u}) \stackrel{c}{=} \sum_{j=1}^{J} \frac{\partial \mathcal{J}}{\partial \mathbf{u}^t \mathbf{v}^j}(\mathbf{u}_i) \mathbf{u}^t \mathbf{v}^j. \tag{8.16}$$

Let us apply this general approximation to the shape-compactness term in (8.8), with reference point c assumed fixed. It is easy to see that (8.8) befits the general form of high-order functions in (8.1), and could be written as follows:

$$\mathcal{J}(\mathbf{u}) = g(\mathbf{u}^t \mathbf{v}^1, \mathbf{u}^t \mathbf{v}^2), \tag{8.17}$$

where

$$g(x, y) = \frac{x}{y^2}$$

$$\mathbf{u}^t \mathbf{v}^1 = \sum_{p \in \Omega} v_p^1 u_p; \; v_p^1 = \|p - c\|^2$$

$$\mathbf{u}^t \mathbf{v}^2 = \sum_{p \in \Omega} v_p^2 u_p; \; v_p^2 = 1. \tag{8.18}$$

For the particular case defined by expressions (8.18), we have:

$$\frac{\partial \mathcal{J}}{\partial \mathbf{u}^t \mathbf{v}^1}(\mathbf{u}_i) = \frac{1}{\left(\mathbf{u}_i^t \mathbf{v}^2\right)^2}$$

$$\frac{\partial \mathcal{J}}{\partial \mathbf{u}^t \mathbf{v}^2}(\mathbf{u}_i) = \frac{-2\mathbf{u}_i^t \mathbf{v}^1}{\left(\mathbf{u}_i^t \mathbf{v}^2\right)^3}. \tag{8.19}$$

Plugging these expressions in the general first-order Taylor expansion in (8.16) yields the following unary approximation of the shape compactness term:

$$\mathcal{J}_i(\mathbf{u}) \stackrel{c}{=} \frac{1}{\left(\mathbf{u}_i^t \mathbf{v}^2\right)^2} \mathbf{u}^t \mathbf{v}^1 - \frac{2\mathbf{u}_i^t \mathbf{v}^1}{\left(\mathbf{u}_i^t \mathbf{v}^2\right)^3} \mathbf{u}^t \mathbf{v}^2. \tag{8.20}$$

References

[1] L. Gorelick, F.R. Schmidt, Y. Boykov, Fast trust region for segmentation, in: IEEE Conference on Computer Vision and Pattern Recognition (CVPR), 2013, pp. 1714–1721.
[2] I. Ben Ayed, K. Punithakumar, S. Li, Distribution matching with the Bhattacharyya similarity: A bound optimization framework, IEEE Transactions on Pattern Analysis and Machine Intelligence 37 (9) (2015) 1777–1791.

[3] H. Kervadec, H. Bahig, L. Letourneau-Guillon, J. Dolz, I. Ben Ayed, Beyond pixel-wise supervision: Semantic segmentation with higher-order shape descriptors, in: Medical Imaging with Deep Learning (MIDL), 2021, pp. 1–16.

[4] M. Niethammer, C. Zach, Segmentation with area constraints, Medical Image Analysis 17 (1) (2013) 101–112.

[5] H. Kervadec, J. Dolz, M. Tang, E. Granger, Y. Boykov, I. Ben Ayed, Constrained-CNN losses for weakly supervised segmentation, Medical Image Analysis 54 (2019) 88–99.

[6] Y. Boykov, O. Veksler, R. Zabih, Fast approximate energy minimization via graph cuts, IEEE Transactions on Pattern Analysis and Machine Intelligence 23 (11) (2001) 1222–1239.

[7] Y. Boykov, V. Kolmogorov, An experimental comparison of min-cut/max-flow algorithms for energy minimization in vision, IEEE Transactions on Pattern Analysis and Machine Intelligence 26 (9) (2004) 1124–1137.

[8] Y. Yuan, A review of trust region algorithms for optimization, in: International Congress on Industrial & Applied Mathematics (ICIAM), 1999, pp. 271–282.

[9] S. Boyd, L. Vandenberghe, Convex Optimization, Cambridge University Press, 2004.

[10] Y. Boykov, V. Kolmogorov, D. Cremers, A. Delong, An integral solution to surface evolution PDEs via geo-cuts, in: European Conference on Computer Vision, Springer, 2006, pp. 409–422.

[11] I. Ben Ayed, M. Wang, B. Miles, G.J. Garvin, TRIC: Trust region for invariant compactness and its application to abdominal aorta segmentation, in: Medical Image Computing and Computer-Assisted Intervention (MICCAI), in: LNCS, vol. 8673, 2014, pp. 381–388.

[12] J.D. Zunic, K. Hirota, P.L. Rosin, A Hu moment invariant as a shape circularity measure, Pattern Recognition 43 (1) (2010) 47–57.

[13] Y. Boykov, G. Funka Lea, Graph cuts and efficient n-d image segmentation, International Journal of Computer Vision 70 (2) (2006) 109–131.

CHAPTER 9

Random field losses for deep networks

9.1 Fully supervised segmentation

Recently, deep learning techniques have emerged as a powerful classification tool, achieving ground-breaking performances in a breadth of pattern recognition problems [1]. In particular, supervised convolutional neural networks (CNNs) are dominating almost all aspects of computer vision, in problems such as object detection, action recognition, and semantic segmentation, among many others. Given a very large amount of annotated (labeled) training data and ever-growing hardware capabilities, these supervised learning techniques can effectively model semantic objects. This has triggered wide academic and industrial interest in adapting these supervised techniques to many large-scale applications. CNNs are very flexible because they alleviate the need for building task-specific features, which makes them useful in a wide range of image analysis problems. They are trained end-to-end to learn a hierarchy of image features representing various levels of abstraction. The deeper one goes into the hierarchy, the higher the level of the representation [2]. In contrast to conventional classifiers based on hand-crafted features, CNNs can learn both the features and classifier simultaneously, in a data-driven manner. In semantic segmentation problems, CNNs are ubiquitous in the recent literature, achieving ground-breaking levels of performance when full-supervision is available, in a breadth of computer vision and medical-imaging applications [3–5].

Let $\mathbf{X} = (X_p)_{p\in\Omega}$ be a set of random variables that describe a pixelwise labeling of an image $\mathbf{f} = (f_p)_{p\in\Omega}$, with $\Omega \subset \mathbb{R}^{2,3}$ the spatial image domain. The possible values of each X_p is a finite set of discrete labels $\{0, \ldots, L-1\}$. In semantic segmentation of color images, for instance, each label describes a semantic category, e.g., "car", "sky", "bicycle", etc. A labeling is a particular assignment $\mathbf{x} = (x_p)_{p\in\Omega}$ to $\mathbf{X}$ that assigns a label $x_p \in \{0, \ldots, L-1\}$ to pixel p. Standard supervised CNN segmentation uses a parametric probability distribution in the form of a product of posteriors:

$$\prod_p s_p(x_p = l|\boldsymbol{\theta}, \mathbf{f}),$$

High-Order Models in Semantic Image Segmentation
https://doi.org/10.1016/B978-0-12-805320-1.00014-2

with $\boldsymbol{\theta}$ a vector containing the network parameters. Typically, each posterior is assumed to follow a softmax form:

$$s_p(x_p = l|\boldsymbol{\theta}, \mathbf{f}) = \frac{1}{R_p} \exp r_p(l; \boldsymbol{\theta}, \mathbf{f}), \qquad (9.1)$$

where $R_p = \sum_{l=0}^{L-1} \exp r_p(l; \boldsymbol{\theta}, \mathbf{f})$ is a normalization constant and r_p a real scalar function representing the output of the network for pixel p. For notation simplicity, we will omit the dependence of s_p on $\boldsymbol{\theta}$ and $\mathbf{f}$ as this does not result in any ambiguity in the presentation. We will simply use $s_{p,l}$ to denote $s_p(x_p = l|\boldsymbol{\theta}, \mathbf{f})$. In the standard fully supervised setting, and for each training image, the problem amounts to minimizing with respect to parameters $\boldsymbol{\theta}$ a loss of the following form:

$$\sum_{p\in\Omega} \mathcal{H}(\mathbf{z}_p, \mathbf{s}_p), \qquad (9.2)$$

where $\mathbf{z}_p \in \{0, 1\}^L$ is a binary vector indicating the ground-truth label of point p: $\mathbf{z}_p = (z_{p,0}, \dots, z_{p,L-1})^t$, with $z_{p,l} = 1$ if pixel p has label l and $z_{p,l} = 0$ otherwise. Vector $\mathbf{s}_p = (s_{p,0}, \dots, s_{p,L-1})^t \in [0, 1]^L$ contains the softmax probability outputs of the network for pixel p. Both $\mathbf{z}_p$ and $\mathbf{s}_p$ are in the L-dimensional probability simplex, which we denote: $\nabla_L = \{\mathbf{y} \in [0, 1]^L \mid \mathbf{1}^t\mathbf{y} = 1\}$. $\mathcal{H}$ measures some divergence between the output probabilities of the network and ground-truth labels. A typical choice is the standard cross-entropy:

$$\mathcal{H}(\mathbf{z}_p, \mathbf{s}_p) = -\sum_l z_{p,l} \ln s_{p,l}. \qquad (9.3)$$

Training the network, i.e., optimization with respect to parameters $\boldsymbol{\theta}$, is typically carried out with standard stochastic gradient descent (SGD) [1]. This amounts to updating network parameters $\boldsymbol{\theta}$ using gradient steps, each based on subsets (mini-batches) of training samples. Gradient computations for the parameters within each layer of the network are based on standard back-propagation. The purpose of this chapter is not to discuss the standard aspects of training deep CNNs such as architecture or parameter-optimization choices. In fact, the literature abounds in resources on the subject since CNNs have become very popular in computer vision. Rather, our purpose here is to discuss CNN segmentation in the weakly supervised setting, in which case the pairwise or high-order models discussed in this book are shown to be very helpful [6–11]. In fact, the main drawback of fully supervised CNN models is that they require a large amount of training data. For instance, in the context of semantic segmentation, full supervision requires laborious and costly annotations of all the pixels in the image, which may not be available in a breadth of applications, and even more so when dealing with volumetric data, as in medical imaging, or with massive amounts of high-resolution video data. So far, this has limited the viability of fully supervised learning techniques in many applications, and has recently triggered much research interest in weakly supervised learning methods that do not require full annotations and scale up to large problems and data sets.

9.2 Weakly supervised segmentation

Weakly supervised learning is a rapidly evolving area of machine learning. It extends various types of learning paradigms, for instance, *semi-supervised learning (SSL)*, in which we use a mixture of labeled and unlabeled data points (or samples), or *multiple-instance learning (MIL)*,[1] in which we learn from labels at the level of sets (or groups) of samples. In semantic segmentation, weak supervision with partial and/or set-level labels may take various forms; see the examples in Fig. 9.1. For example, scribbles [6,7,9] or points [13] within the target semantic regions can be viewed as a form of SSL, in which only a fraction of the pixels of a training image is annotated. Image tags[2] [11,14] or bounding boxes enclosing the target semantic regions [8,15] can be viewed as a form of MIL.

Imposing prior knowledge on the probability outputs of a deep network can mitigate the lack/uncertainty of annotations and is a well-established general principle in machine learning [1,16]. Acting as regularizers, such priors leverage unlabeled samples with domain-specific knowledge. In semantic segmentation, several recent works showed that adding *unsupervised* loss terms, such as pairwise conditional random fields (CRFs) [7,17], high-order balanced graph clustering [6] or priors on the sizes of the target

---

[1] MIL uses training samples grouped into sets, commonly referred to as *bags* [12]. Supervision is provided only for entire sets—a single positive or negative label is assigned to each bag. A positive label for a given class and bag of samples indicates that the bag includes at least one sample belonging to the class, while a negative label indicates that all the instances of the bag do not belong to the class.

[2] Image tags are a form of weak supervision at the image level, not the pixel level. In this case, the supervision information is whether a semantic target region, e.g., a car, is present or absent in a given training image [11].

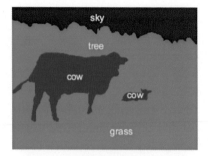

| Image | Full supervision |

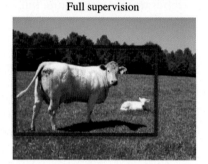

| Scribble supervision (semi-supervised) | box supervision (MIL) |

Figure 9.1 Full and weak supervision for semantic segmentation. The first row shows a training image on the left side and full supervision (annotations of all the pixels in the image) on the right. Manually assigning a semantic label to each pixel is a time-consuming task, more so when the number of training images is very large. The second row shows two different forms of weak supervision. The left side depicts scribble-based annotations, which can be viewed as a form of semi-supervision, in which only a fraction of the pixels of a trained image is annotated. The left side depicts a box annotation for the class "cow". This can be viewed as a form of MIL supervision: the set (bag) of pixels inside the box has positive labels for the class "cow" (indicating that the set includes at least one pixel belonging to the class) and the set of pixels outside the box has a negative label (none of them belongs to the class "cow"). Another very similar form of MIL supervision is image tags, which are image-level labels. For example, for this particular image, classes "cow", "sky" and "tree" are positive and classes "person", "road" and "table" are negative. (*Figures from [9] © IEEE.*)

regions [10,18,19], can achieve outstanding levels of performance, close to full-supervision results, but using only fractions of the ground-truth labels.

9.2.1 Regularized losses

Let $\Omega_{\mathcal{L}} \subset \Omega$ a set of labeled pixels that corresponds to a fraction of the points within training-image domain Ω. For instance, such partial labels can be obtained with few scribbles; see the example in Fig. 9.1. The recent studies in [6,7,17] examined *regularized losses* of the following general form:

$$\min_{\theta} \sum_{p \in \Omega_{\mathcal{L}}} \mathcal{H}(\mathbf{z}_p, \mathbf{s}_p) + \lambda \mathcal{R}(\mathbf{s}). \tag{9.4}$$

The first term in Eq. (9.4) is a *partial* cross-entropy[3] loss. Here, the difference with full supervision is that the summation is over the subset of pixels that are labeled, not over the whole image domain. We use the same notation for the ground-truth labels as in the fully supervised setting in Eq. (9.3), i.e., simplex vectors $\mathbf{z}_p \in \{0, 1\}^L$. This does not result in any ambiguity in subsequent discussions. $\mathcal{R}(\mathbf{s})$ is an unsupervised regularization loss that depends on network predictions $s_{p,l}$ for all pixels p in Ω, both unlabeled and labeled. Vector $\mathbf{s} = (\mathbf{s}_p)_{p \in \Omega} \in [0, 1]^{L|\Omega|}$ contains the network predictions for all the pixels in the image. Next, we will discuss in more detail examples of regularization losses.

9.2.1.1 Potts/CRF loss

The outputs of the network can be viewed as *relaxations*[4] (or continuous versions) of binary indicator variables $\mathbf{s}_p = (s_{p,0}, \ldots, s_{p,L-1})^t \in \{0, 1\}^L$, with $s_{p,l} = 1$ if p belongs to region l and $s_{p,l} = 0$ otherwise. In this case of discrete binary variables, the standard Potts/CRF model, which we discussed earlier in this book, can be written in the following form:

$$\mathcal{R}_{\mathrm{CRF}}(\mathbf{s}) = \sum_{p,q \in \Omega} w_{pq} \, [\mathbf{s}_p \neq \mathbf{s}_q] \; = \; \sum_{p,q \in \Omega} w_{pq} \, \|\mathbf{s}_p - \mathbf{s}_q\|^2, \tag{9.5}$$

where, we recall, w_{pq} is a pairwise cost measuring some similarity between pixels p and q, e.g., a monotonically decreasing function of color difference. [.] denotes the Iverson bracket, taking value 1 if its argument is true and 0 otherwise. As suggested in [7] for weakly supervised segmentation, one can use a relaxed version of the Potts regularization in Eq. (9.5), in which

---

[3] We use the cross entropy as an example, as it is commonly used as a loss for segmentation. However, our discussion of unsupervised regularization terms and constraints is not restricted to any specific form of loss for the set of labeled points.

[4] "Relaxation" means that we relax each integer constraint $\mathbf{s}_p \in \{0, 1\}^L$, replacing it with $\mathbf{s}_p \in [0, 1]^L$.

relaxed variables $\mathbf{s}_p \in [0, 1]^L$ are the standard softmax outputs of a CNN. Notice that, for relaxed variables, the second equality in (9.5) does not hold anymore. The quadratic function in the right-hand side of (9.5) can be viewed as a particular relaxation of the discrete-variable Potts model. In fact, such a quadratic function is widely used as an unsupervised loss term in the general context of semi-supervised deep learning [16]. Of course, this choice of relaxation is not unique. The following is another relaxation of discrete Potts [7]:

$$\mathcal{R}_{\text{CRF}}(\mathbf{s}) = - \sum_{p,q \in \Omega} w_{pq} \mathbf{s}_p^t \mathbf{s}_q = - \sum_l (\mathbf{s}^l)^t W \mathbf{s}^l, \tag{9.6}$$

with $\mathbf{s}^l$ a vector of length $|\Omega|$ taking the form $\mathbf{s}^l = (s_{1,l}, \ldots, s_{|\Omega|,l})^t$. For binary variables $s_{p,l} \in \{0, 1\}$, $\mathbf{s}^l$ is an indicator function for region l. $W = [w_{p,q}]$ is the $|\Omega| \times |\Omega|$ matrix of pairwise potentials. Notice that, for binary variables, we can write the Potts model in (9.5) as follows:

$$\sum_{p,q \in \Omega} w_{pq} \|\mathbf{s}_p - \mathbf{s}_q\|^2 = 2 \sum_{p \in \Omega} d_p - 2 \sum_{p,q \in \Omega} w_{pq} \mathbf{s}_p^t \mathbf{s}_q,$$

with $d_p = \sum_q w_{pq}$. Term $\sum_{p \in \Omega} d_p$ is a constant independent of segmentation variables $\mathbf{s}_p$. In the case of binary variables, minimizing (9.6) is equivalent to minimizing the function on the right-hand side of (9.5), up to an additive constant. The relaxation in (9.6) can be written using the same form we gave in Chapter 3 for the popular Dense CRF model [20]:

$$\mathcal{R}_{\text{CRF}}(\mathbf{s}) = - \sum_{p,q \in \Omega} w_{pq} \mathbf{s}_p^t \mathbf{s}_q = -\mathbf{s}^t \Psi \mathbf{s},$$

where Ψ is the Kronecker product between W and the $L \times L$ identity. Notice that this relaxation is concave with respect to $\mathbf{s}$, whereas the relaxation on the right-hand side of (9.5) is convex.

There are various choices of pairwise affinity matrix W. It can be a sparse CRF as in the popular graph-cut techniques [21,22] discussed earlier in this book. Another choice, which we also discussed in great detail in Chapter 3, is the Dense CRF model in [20]. In this case, affinity matrix W, which is computed from fully connected Gaussian kernels, is dense. Therefore, a naive implementation of back-propagation for the regularization loss in Eq. (9.6) would be computationally intractable, with a quadratic complexity in the number of pixels in the image ($O(|\Omega|^2)$). In fact, back-propagation requires computation of the gradient of such dense pairwise terms, which takes the following form in the case of (9.6):

$$\frac{\partial \mathcal{R}_{\mathrm{CRF}}(\mathbf{s})}{\partial \mathbf{s}^l} = -2W\mathbf{s}^l. \tag{9.7}$$

In Chapter 3, we discussed how fully connected graphs can be tackled efficiently via fast Gaussian filtering in the context of mean-field inference [20,23]. The same fast Gaussian filtering principle, which is common in signal processing [24], can be used here for computing gradient (9.7). This reduces significantly the computational burden from quadratic to linear.

9.2.1.2 Normalized cut loss

In Chapter 6, we discussed in great details ratio functions for graph clustering, e.g., the very popular NC objective:

$$\mathcal{R}_{\mathrm{NC}}(\mathbf{s}) = -\sum_l \frac{(\mathbf{s}^l)^t W \mathbf{s}^l}{\mathbf{d}^t \mathbf{s}^l}, \tag{9.8}$$

where, as earlier in the case of CRF, $\mathbf{s}^l$ is a binary indicator vector for cluster (or region) l and $W = [w_{pq}]$ is an $|\Omega| \times |\Omega|$ matrix of pairwise potentials w_{pq}, each evaluating an affinity (similarity) between points p and q, e.g., via some kernel function. Vector $\mathbf{d} = (d_p)_{p \in \Omega}$ contains point degrees: $d_p = \sum_q w_{pq}$. Within the learning community, graph-clustering objectives such as NC are widely used for partitioning high-dimensional features [25,26], and standard spectral relaxation [27,28] is a dominant technique for optimizing such discrete-variable objectives. Similarly to the case of CRF, one can use a relaxed version of NC objective (9.8) in weakly supervised segmentation, where variables $\mathbf{s}_p \in [0, 1]^L$ becomes the standard softmax outputs of a CNN [6]. Of course, in this case, we do not use spectral relaxation. A standard gradient descent approach can be used for training a CNN with the NC loss in Eq. (9.8). Similarly to before, back-propagation requires computation of the gradient of the NC objective, which is given by:

$$\frac{\partial \mathcal{R}_{\mathrm{NC}}}{\partial \mathbf{s}^l} = \frac{(\mathbf{s}^l)^t W \mathbf{s}^l \mathbf{d}}{(\mathbf{d}^t \mathbf{s}^l)^2} - 2\frac{W\mathbf{s}^l}{\mathbf{d}^t \mathbf{s}^l}. \tag{9.9}$$

Again, a naive implementation of back-propagation with this gradient expression would be of intractable quadratic complexity. Fortunately, we can use fast Gaussian filtering [24], as in the case of the CRF loss, to reduce the computational burden from quadratic to linear.

In Chapter 6, we also discussed the KC model [26,29], showing that integrating the NC term and a CRF regularization in a single model can be powerful in the context of classical interactive segmentation problems. KC combines the complementary benefits of balanced NC partitioning and region boundary regularization or edge alignment as in the Potts model:

$$\mathcal{R}_{KC}(\mathbf{s}) = \mathcal{R}_{CRF}(\mathbf{s}) + \gamma \mathcal{R}_{NC}(\mathbf{s}), \tag{9.10}$$

where γ is a positive constant to weigh the contribution of each term. Again, one can use a relaxed version of this KC objective in weakly supervised segmentation [7], where variables $\mathbf{s}_p \in [0, 1]^L$ become CNN probability outputs.[5] This unsupervised KC loss achieved a state-of-the-art performance in weakly supervised semantic segmentation of color images, approaching full-supervision accuracy [7]. In the following, we will discuss in more detail related experimental results.

9.2.1.3 Discussion of experimental results

The study in [7] reported comprehensive evaluations of regularized losses in the context of scribble-based semantic segmentation [9], using the PASCAL VOC12 data set [30]. The data set contains 10,582 images for training and 1,449 images for validation, along with ground-truth segmentations. The weak annotations, which were introduced in [9], are partial labels taking the form of scribbles for each training image; see the example in Fig. 9.1. Such a partial annotation is provided for only 3% of the training data, i.e., for the fraction of pixels corresponding to the scribbles. The implementation is based on DeepLab-ResNet101 [31], a state-of-the-art network for fully supervised semantic segmentation. A dense Gaussian kernel defined over color and spatial-position features is used for both $\mathcal{R}_{CRF}$, $\mathcal{R}_{NC}$ and $\mathcal{R}_{KC}$, with a fast Gaussian filtering [24] that reduces the computational burden of gradient computation for dense pairwise potentials from quadratic to linear. A common measure for evaluating semantic segmentation of color images is the Intersection-Over-Union (IoU), which assesses a similarity between an obtained segmentation region $S \subset \Omega$ and a ground-truth segmentation

---

[5] In it worth noting that, in the context of classical segmentation techniques, integrating NC and CRF objectives, as in the KC model, is not straightforward, due to significant differences in the standard optimization techniques used for each of these objectives (e.g., spectral relaxation for NC and graph cuts or mean-field inference for CRF) [26,29]. However, the relaxed version of KC used as unsupervised loss for deep networks can be directly tackled with gradient descent (if the chosen relaxation is differentiable).

Table 9.1 Evaluations of the mean IoU on the PASCAL VOC2012 validation set. The values within the parentheses indicate the gap with respect to full-supervision performance. With regularization, one can reach above 96% of the full-supervision performance with only 3% of the pixels labeled (i.e., the pixels corresponding to the scribbles). The results are reported from [7].

Weak supervision (scribble-based)				Full
Partial cross-entropy only	w/ NC [6]	w/ CRF [7]	w/ KC [7]	
69.5 (6.1)	72.8 (2.8)	72.9 (2.7)	73.0 (2.6)	75.6

region $G \subset \Omega$: $IoU = 100\frac{|S \cap G|}{|S \cup G|}$, where $|.|$ denotes cardinality. The higher the IoU, the better the result. Table 9.1 reports the mean IoU over the validation set for each regularization loss. It also reports the validation-set performances of full supervision and training with partial annotations only (i.e., optimizing the partial cross-entropy without any regularization loss). For each model, the weights of the regularization losses are chosen to achieve the best accuracy on the validation set. The benefit of the regularization losses is clear, with the best performance obtained with the KC loss, which reached over 96% of the full-supervision performance. Fig. 9.2 depicts a few representative examples of the results, illustrating the benefit of adding regularization losses to the partial cross-entropy. Notice that the CRF loss has an effect different from the NC loss, with the former producing better edge alignments and the latter yielding color clustering with more balanced segmentation regions. The KC loss integrates the benefits of both, yielding the best results, both visually and quantitatively.

9.2.1.4 On Grid CRF and gradient descent

In the experiments discussed earlier, we examined a CRF loss of the form (9.5), with dense neighborhoods, a model often referred to as *Dense CRF* [20]. In this case, affinity matrix W is computed from fully connected Gaussian kernels. It is possible, however, to use a sparse form of this regularization, often referred to as *Grid CRF* [17], where only neighboring pixels are connected. This is the case in the standard graph–cut segmentation techniques we discussed in Chapter 1. Recall that Grid Potts/CRF model optimizes a contrast-weighted, region–boundary length, thereby encouraging smooth boundaries that are well aligned with the image edges. In fact, in classical, nondeep segmentation methods, Grid Potts/CRF models are very popular due to their robustness, geometrical motivations and amenability to powerful discrete graph–cut solvers, which guarantee global

Figure 9.2 Regularized losses for weakly supervised semantic segmentation. First column: the image; second column: partial cross-entropy only; third column: NC loss; fourth column: CRF loss; fifth column: KC loss; sixth column: ground truth. (*Figures from [7] © Springer.*)

optima for binary objectives [21] and some solution-quality bounds for multi-label objectives [32]. In those standard methods, typically built upon hand-crafted color classifiers such as those we discussed in Chapter 1, Grid CRF is generally preferred over Dense CRF. The latter has a weaker regularization effect [33], typically yielding irregular (noisy) boundaries [20]. The nondeep segmentation examples in Fig. 9.3 depict an illustration of this. In some instances, Dense CRF may have the desirable effect of preserving thin structures [33], which could be over-smoothed by sparse length regularization [31]. However, this is due to the same weak-regularization effect preserving boundary noise in Fig. 9.3 (c) [17]. Another important limitation of Dense CRF is that well-known and global discrete optimization techniques, such as graph cuts [21], do not scale to dense connections.

These several advantages of Grid CRF over Dense CRF, well-established in classical computer vision, do not materialize in the context of modern deep CNNs. In fact, dense CRF models are the de facto choice in the context of CNNs [5,7], often achieving better performances than Grid CRFs. For instance, replacing the dense connections by sparse ones in the PASCAL VOC2012 experiments we discussed earlier, and keeping exactly

| (a) image + scribbles | (b) grid CRF [21] | (c) dense CRF [20] |

Figure 9.3 Nondeep segmentation example. Grid (sparse) CRF, optimized via a graph cut, yields smoother segmentation boundaries and better alignment with edges than Dense CRF with first-order optimization (mean-field inference). *(Figures from [17] © IEEE.)*

the same settings,[6] the mean IoU drops from 72.9% for Dense CRF to 71.7% for Grid CRF. With those CNN segmentation performances, which are inconsistent with the nondeep segmentation results in Fig. 9.3, one may think that Dense CRF is a better regularization loss than Grid CRF. In fact, and as we will discuss later in Section 9.3, the issue is in the optimization technique, not in the loss function and its regularization properties. Gradient descent is the default method for training neural networks, as it enables to optimize any differentiable regularizer. However, it is well known in classical, MRF-based computer vision that gradient descent is not a good optimization technique for many powerful regularization functions, such as Grid CRF [17].

Let us take the simple 1D–image example in Fig. 9.4, introduced in [17]. This example gives some insight as to why gradient descent, or other first-order approximation techniques, may have difficulty with the Grid CRF loss, although they handle well Dense CRF. For each 1D pixel coordinate t, the example generates a two–region segmentation of the 1D image in Fig. 9.4 (a) as follows. The foreground is given by $S^1(t) = \{p \in \Omega | p < t\} = \{p \in \Omega | s_{p,1} = 1\}$, and the background is its complement, given by $S^0(t) = \Omega \setminus S^0(t)$. Figs. 9.4 (b) and (c) plot the CRF losses as functions of the values of t, each corresponding to a different segmentation of the 1D image. Clearly, the dense connections yield a "smoother" objective function that may facilitate gradient-based optimization. Note that the "flatter" minimum for Dense CRF makes detecting the discontinuities (i.e., the image edges) more difficult, which might explain in part the noisy boundary obtained by Dense CRF for the 2D real example in Fig. 9.3 (c).

---

[6] By "same setting" we mean the choices of the deep–network architecture and optimization technique.

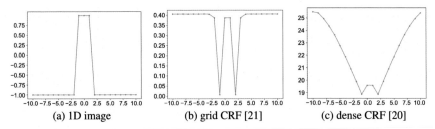

(a) 1D image   (b) grid CRF [21]   (c) dense CRF [20]

Figure 9.4 Synthetic 1D-image example. Plots of the Grid and Dense CRF terms as functions of several segmentations, each corresponding to a pixel coordinate.

The simple synthetic example in Fig. 9.4 suggests that gradient-based learning methods may not be appropriate for sparse regularization losses, which may have many local minima. In Section 9.3, we will discuss an alternative to gradient descent for training deep networks with the Grid CRF loss, which was introduced in [17]. The general framework in [17] enables to take advantage of well-known and effective discrete graph-cut solvers, which we discussed in great detail in Chapter 2. We will discuss a network training that uses a discrete α-expansion solver, which, for Grid CRF loss, improves significantly optimization quality in comparison to gradient descent.

9.2.2 Segmentation proposals

Proposal-based methods are weakly supervised semantic segmentation techniques, which mimic full supervision by generating full training labels (segmentation proposals) using only partial annotations. The proposals can be viewed as synthesized ground-truth labels, which are used to train a CNN. In general, these techniques, which are very common in the recent literature [8,9,11,14,15,34], follow an iterative process that alternates two steps: (i) standard SGD for training a CNN from the proposals and (ii) standard regularization-based segmentation, which yields the proposals. This second step typically uses one of the standard techniques that we examined in great detail earlier in the book, i.e., mean-field inference [8,14] or graph cuts [9]. In particular, the dense CRF regularizer of Krähenbühl and Koltun [20], facilitated by fast mean-field inference, has become very popular in semantic CNN segmentation, both in the fully [5,35] and weakly [8,14,34] supervised settings. This followed from the great success of DeepLab [35], which popularized the use of dense CRF and mean-field inference as a postprocessing step in the context of fully supervised CNN segmentation.

In the following, we examine how proposal techniques based on mean-field inference can be interpreted as an alternating direction method (ADM) [36] for optimizing regularization loss (9.6), a connection established recently in the work in [7].

9.2.2.1 Connection to regularization losses

Proposal-based methods alternate two steps, one corresponding to training a CNN[7] and the other to computing a segmentation proposal. Let vector $\mathbf{u} = (\mathbf{u}_p)_{p \in \Omega} \in \{0, 1\}^{L|\Omega|}$ describes a segmentation proposal obtained from the previous iteration. It concatenates binary vectors $\mathbf{u}_p = (u_{p,0}, \ldots, u_{p,L-1})^t \in \{0, 1\}^L$, each being within the L-dimensional probability simplex: $u_{p,l} = 1$ if pixel p has label l and $u_{p,l} = 0$ otherwise. For each unlabeled pixel $p \in \Omega_{\mathcal{U}} \subset \Omega$, $\mathbf{u}_p$ denotes the synthesized ground-truth (proposal computed at the previous iteration) and, for labeled pixels $p \in \Omega_{\mathcal{L}} \subset \Omega$, $\mathbf{u}_p$ is fixed and constrained to be equal to the available annotation $\mathbf{z}_p$. With ground-truth proposal $\mathbf{u}$ fixed (computed at the previous iteration), the training step learns the network parameters $\boldsymbol{\theta}$ by optimizing the following loss via standard SGD and back-propagation techniques:

$$\tilde{\boldsymbol{\theta}} = \arg \min_{\boldsymbol{\theta}} \sum_{p \in \Omega_{\mathcal{L}}} \mathcal{H}(\mathbf{z}_p, \mathbf{s}_p) + \sum_{p \in \Omega_{\mathcal{U}}} \mathcal{H}(\mathbf{u}_p, \mathbf{s}_p). \tag{9.11}$$

Given fixed outputs of the network $\tilde{\mathbf{s}} = (\tilde{\mathbf{s}}_p)_{p \in \Omega}$, with $\tilde{\mathbf{s}}_p = (\tilde{s}_{p,0}, \ldots, \tilde{s}_{p,L-1})^t$ and $\tilde{s}_{p,l} = \frac{1}{R_p} \exp r_p(l; \tilde{\boldsymbol{\theta}}, \mathbf{f})$, the second step computes the next segmentation proposal. This step is based on optimizing a standard regularization function containing unary and pairwise potentials, i.e., taking a from similar to the basic segmentation functions we discussed in Chapter 1:

$$\min_{\mathbf{u} \in \{0,1\}^{L|\Omega|}} \sum_{p \in \Omega_{\mathcal{U}}} \mathcal{H}(\mathbf{u}_p, \tilde{\mathbf{s}}_p) + \lambda \mathcal{R}(\mathbf{u}). \tag{9.12}$$

Notice that, when network outputs $\tilde{\mathbf{s}}$ are fixed in Eq. (9.12), each term $u_{p,l} \ln \tilde{s}_{p,l}$ in cross-entropy $\mathcal{H}(\mathbf{u}_p, \tilde{\mathbf{s}}_p)$ is a unary potential for variable $u_{p,l}$. Recall from Chapter 3 that, when regularization term $\mathcal{R}(\mathbf{u})$ takes the form of a dense CRF, then discrete problem (9.12) can be solved efficiently with parallel mean-field inference [20,23,37], which is further accelerated with

---

[7] Training a CNN from proposals does not need to continue until convergence; switching from learning network parameters to proposal generation can occur during the substeps of training of one CNN.

fast Gaussian filtering techniques [24]. As discussed in great detail in Chapter 3, the main idea of mean-field algorithms is to approximate the Gibbs distribution corresponding to a discrete function by a simpler factorial distribution via minimizing the KL divergence. One can easily show that, for pairwise dense CRF regularization, mean-field approximation of discrete problem (9.12) corresponds to solving the following problem subject to simplex constraints $\mathbf{u}_p \in \nabla_L \; \forall p$:

$$\min_{\mathbf{u} \in [0,1]^{L|\Omega|}} \sum_{p \in \Omega_u} \mathcal{H}(\mathbf{u}_p, \tilde{\mathbf{s}}_p) + \lambda \mathcal{R}(\mathbf{u}) - \sum_{p \in \Omega_u} \mathcal{H}(\mathbf{u}_p), \qquad (9.13)$$

where $\mathcal{H}(\mathbf{u}_p)$ is the entropy of probability-simplex variable $\mathbf{u}_p$: $\sum_l u_{p,l} \ln u_{p,l}$. In addition to the mean-field approximation motivation perspective, the problem in (9.13) can be viewed as a direct relaxation of discrete problem (9.12). Notice that the first two terms in (9.13) correspond exactly to problem (9.12), except that integer ("hard") constraints on assignment variables $u_{p,l} \in \{0, 1\}$ are relaxed, i.e., replaced by "soft" constraints $u_{p,l} \in [0, 1]$. Also, the negative entropy (the third term), which comes from the KL divergence of the mean-field approximation, can be interpreted as a barrier (or penalty) function for the simplex constraints on "soft" label-assignment variables $\mathbf{u}_p$. Each term in the sum penalizes deviation of variable $\mathbf{u}_p$ from the "middle" of the simplex, encouraging "softness" of the label assignment. Notice that, for variables at the vertices of the simplex (i.e., hard binary assignments $\mathbf{u}_p$), this negative entropy vanishes, and the relaxation in (9.13) becomes equivalent to discrete problem (9.12). Recall from Chapter 3 that solving a CRF model in conjunction with negative entropy, as in (9.13), can be done by parallel (independent), closed-form updates of variables $\mathbf{u}_p$, while guaranteeing convergence when some conditions are satisfied.[8]

Now let us consider the following proposition, which connects proposal methods to direct regularized losses of the general form in Eq. (9.4).

Proposition 1. *Alternating optimization of* (9.11) *and* (9.13) *can be interpreted as an alternating direction method (ADM),*[9] *which optimizes loss* (9.4) *by*

---

[8] Independent updates converge when the CRF model is concave, as is the case of the standard Potts model [37].

[9] The most basic forms of alternating direction methods [36] replace unconstrained problem of $\min_x f(x) + g(x)$ by equality-constrained problem $\min_{x,y} f(x) + g(y)$ s.t $x = y$ and alternates optimization over x and y.

Table 9.2 Evaluations of the mean IoU on the PASCAL VOC2012 validation set for three different versions of proposal generation.

Proposal generation		Direct loss (Dense CRF)
GrabCut (One time)	Dense CRF (ADM)	
63.9	72.5	72.9

decomposing the problem as follows:

$$\min_{\boldsymbol{\theta},\; \mathbf{u}\in[0,1]^{L|\Omega|}} \sum_{p\in\Omega_{\mathcal{L}}} \mathcal{H}(\mathbf{z}_p, \mathbf{s}_p) + \lambda\mathcal{R}(\mathbf{u}) + \sum_{p\in\Omega_{\mathcal{U}}} \mathrm{KL}(\mathbf{u}_p||\mathbf{s}_p), \tag{9.14}$$

where KL denotes the Kullback–Leibler divergence.

Proof. The connection between problem (9.14) and the corresponding subproblems (9.11) and (9.13) follows directly from expressing the KL divergence in term of the entropies: $\mathrm{KL}(\mathbf{u}_p|\mathbf{s}_p) = \mathcal{H}(\mathbf{u}_p, \mathbf{s}_p) - \mathcal{H}(\mathbf{u}_p)$. □

Therefore, rather that optimizing (9.4) directly with respect to the parameters of the network, proposal-based methods replace network outputs in regularization term $\mathcal{R}(\mathbf{s})$ by a new latent variable $\mathbf{u}$ (the proposal), and use the KL divergence as a penalty for equality constraint $\mathbf{s} = \mathbf{u}$, thereby splitting the problem into two subproblems. This is very similar to the principle of the alternating-direction method of multipliers (ADMM) [36], although there are a few differences. In (9.14), equality constraint $\mathbf{s} = \mathbf{u}$ is handled with a penalty, unlike the standard form of ADMM, in which we use a multiplier penalty (or augmented Lagrangian) to deal with the constraint. Another difference is that the splitting is not performed directly with respect to problem variables, i.e., in this case, network parameters $\boldsymbol{\theta}$. It is performed with respect to the outputs of the network.

9.2.2.2 Proposals vs. direct Dense CRF loss

Table 9.2 compares the performances of different versions of proposal generation and the direct regularized loss approach we examined earlier, for Dense CRF, using the same PASCAL VOC12 data set. For instance, one can use GrabCut to generate a "fake" ground truth that is used subsequently for training a CNN, as in the fully supervised setting. From the results, it is clear that this one-time proposal generation yields much poorer performance than alternating in an iterative manner proposal generation and CNN training, as in ADM. As discussed earlier, for Dense CRF loss

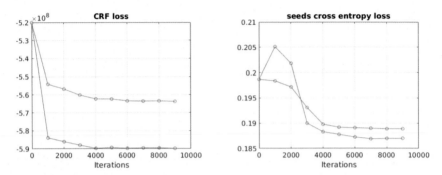

Figure 9.5 Optimization-quality comparison. Evolution of the losses (Dense CRF and partial cross-entropy) for direct SGD optimization (blue) and ADM proposals (red) with the mean-field inference solver. (*Figures from [7]* © *Springer.*)

(9.6), iterative proposal-generation schemes, which alternate CNN training and mean-field inference, can be viewed as an approximate ADM optimization of the loss. While ADM improves the performance of one-time proposal generation, the mean IoU values in Table 9.2 suggest that direct SGD optimization of Dense CRF loss (9.6) yields better performance than ADM splitting, while being more efficient (since it avoids completely proposal generation). One can also compare direct SGD optimization and proposal generation (ADM) from the optimization-quality perspective, as both are optimization of the same Dense CRF loss. Fig. 9.5 juxtaposes the two optimization strategies in terms of the loss values obtained during training. ADM minimization, which alternates SGD training and proposal generation, yielded higher values of the losses than direct, computationally more efficient SGD optimization of Dense CRF. This may be due to two facts. First, mean-field inference is a first-order optimization method, similar to gradient descent. Therefore, in this case, and as shown by the plots in Fig. 9.5, ADM does not bring a gain in terms of optimization quality in comparison to direct SGD optimization of Dense CRF. This might explain in part the lower accuracy obtained by ADM (Table 9.2). Second, proposal generation can be viewed as an approximate (not exact) ADMM optimization of the loss: As mentioned earlier, the splitting in ADM is not performed directly with respect to problem variables, i.e., network parameters θ. It is performed with respect to the outputs of the network.

9.3 Beyond gradient descent for random field losses

Gradient descent is the de facto method for training deep neural networks. Generally, differentiable loss functions are defined so as to accommodate gradient descent. Therefore, the accuracies obtained from regularized-loss models in deep learning might be limited by the optimization performance of gradient descent. In fact, in the context of classical (nondeep) image segmentation methods, it is well-known that gradient descent may lead to weak local minima for certain regularization functions, e.g., the standard Grid CRF term [17]. The simple synthetic example in Fig. 9.4, which we discussed in detail earlier, suggests that gradient descent may not be a good optimizer for the Grid CRF loss. In fact, the optimization performance of gradient descent might obfuscate the good regularization properties of Grid MRF. Can we take advantage of powerful discrete graph-cut solvers, well known for their effectiveness for submodular and sparse pairwise potentials, and which we discussed in great detail earlier in the book? The recent work in [17] discussed a general optimization method, which can take advantage of effective discrete solvers. Let us revisit the ADM optimization framework we discussed earlier in Proposition 1. In this framework, for fixed network parameters $\boldsymbol{\theta}$, optimization with respect to latent assignment variable $\mathbf{u} \in [0, 1]^{L|\Omega|}$ corresponds to solving:

$$\min_{\boldsymbol{\theta}, \ \mathbf{u} \in [0,1]^{L|\Omega|}} \sum_{p \in \Omega_{\mathcal{U}}} \left[\mathcal{H}(\mathbf{u}_p, \mathbf{s}_p) - \mathcal{H}(\mathbf{u}_p) \right] + \lambda \mathcal{R}(\mathbf{u}). \qquad (9.15)$$

Notice that, for discrete binary variables $\mathbf{u} \in \{0, 1\}^{L|\Omega|}$, entropy $\mathcal{H}(\mathbf{u}_p)$ is null and cross-entropy $\mathcal{H}(\mathbf{u}_p, \mathbf{s}_p) = \sum_l u_{p,l} \ln s_{p,l}$ is a unary potential. Therefore, for a submodular and sparse pairwise term $\mathcal{R}(\mathbf{u})$, we can tackle the subproblem in (9.15) with the powerful discrete α-expansion solver in [32], which we discussed in detail earlier in this book. Indeed, the authors of [17] observed that, for Grid CRF, a deep-network training based on this ADM approach with discrete α-expansion solver improves significantly optimization quality in comparison to gradient descent.

The plot in Fig. 9.6 depicts the Grid CRF loss as a function of the number of training iterations for both gradient descent and ADM with α-expansion solver, confirming the superiority of the latter, which achieves a better (lower) minimum while improving convergence. Notice that, within the first 1,000 training iterations, ADM with α-expansion yielded a grid CRF loss lower than the one obtained by gradient descent at convergence. In this case, gradient descent took about 6,000 training iterations to con-

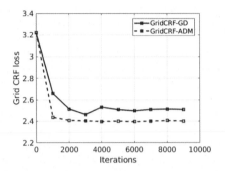

Figure 9.6 Grid CRF loss evolution during training iterations for ADM (with discrete α-expansion solver) and gradient descent. With α-expansion, ADM outperforms significantly gradient descent (GD), yielding a lower training loss and improving convergence. (*Figure from [17] © IEEE.*)

Table 9.3 Evaluations of the mean IoU on the PASCAL VOC2012 validation set for different optimizers (ADM and gradient descent) and different CRF models (Grid and Dense). The results are reported from [17].

Grid CRF		Dense CRF	
Gradient descent	**ADM (α-expansion)**	**Gradient descent**	**ADM (first-order)**
71.7	72.8	72.9	72.5

verge. Note, however, within each iteration, ADM requires running the α-expansion solver [32], which is slower than gradient descent updates.

In this case of a Grid CRF loss, it is clear that gradient descent yielded a poor local minimum. This contrasts with the experiment we discussed in Section 9.2.2.2, which showed that ADM with mean-field inference performed worse than gradient descent. Clearly, the main difference comes from the quality of the discrete α-expansion solver for the problem in (9.15). As discussed in Section 9.2.2.2, mean-field inference is a first-order approximation method, like gradient descent. Therefore, unlike the discrete α-expansion solver, it did not bring improvement of the ADM setting.

The improvement that α-expansion brought in optimizing the Grid CRF loss translates into a better segmentation quality, in term of mean IoU on the PASCAL VOC2012 validation set. With the ADM approach and α-expansion, the Grid CRF loss yielded segmentations much better than when using gradient descent as optimizer, both quantitatively (Table 9.3) and qualitatively (Fig. 9.7).

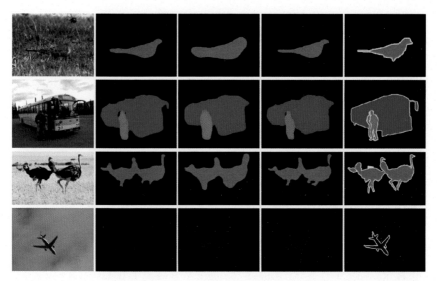

Figure 9.7 The effect of the choice of the optimization technique on the performance of Grid CRF. First column: the image; second column: Dense CRF with gradient descent; third column: Grid CRF with gradient descent; fourth column: Grid CRF with ADM and α-expansion; fifth column: ground-truth. Notice that, with gradient descent, Grid CRF yielded poor, over-smoothed segmentation boundaries (third column). ADM and α-expansion corrected the results of Grid CRF (fourth column), yielding a level of performance comparable to Dense CRF (second column). (*Figures from [7] © IEEE.*)

These results for Grid CRFs suggest that, more generally, when imposing priors on the outputs of CNNs, one may need to pay more attention to the optimization techniques, beyond basic gradient descent. This might be the case for a breadth of discrete priors, which could be amenable to powerful discrete solvers. In fact, the ADM framework in Proposition 1 is quite general and enables tackling any discrete regularizer, as long as there is a good solver for such a regularizer.

References

[1] I. Goodfellow, Y. Bengio, A. Courville, Deep Learning, MIT Press, 2016.
[2] M.D. Zeiler, R. Fergus, Visualizing and understanding convolutional networks, in: European Conference on Computer Vision (ECCV), Part I, 2014, pp. 818–833.
[3] G.J.S. Litjens, T. Kooi, B.E. Bejnordi, A.A.A. Setio, F. Ciompi, M. Ghafoorian, et al., A survey on deep learning in medical image analysis, Medical Image Analysis 42 (2017) 60–88.

[4] J. Long, E. Shelhamer, T. Darrell, Fully convolutional networks for semantic segmentation, in: IEEE Conference on Computer Vision and Pattern Recognition (CVPR), 2015, pp. 3431–3440.

[5] A. Arnab, S. Zheng, S. Jayasumana, B. Romera-Paredes, M. Larsson, A. Kirillov, et al., Conditional random fields meet deep neural networks for semantic segmentation: Combining probabilistic graphical models with deep learning for structured prediction, IEEE Signal Processing Magazine 35 (1) (2018) 37–52.

[6] M. Tang, A. Djelouah, F. Perazzi, Y. Boykov, C. Schroers, Normalized cut loss for weakly-supervised CNN segmentation, in: IEEE conference on Computer Vision and Pattern Recognition (CVPR), 2018, pp. 1818–1827.

[7] M. Tang, F. Perazzi, A. Djelouah, I. Ben Ayed, C. Schroers, Y. Boykov, On regularized losses for weakly-supervised CNN segmentation, in: European Conference on Computer Vision (ECCV), Part XVI, 2018, pp. 524–540.

[8] M. Rajchl, M.C.H. Lee, O. Oktay, K. Kamnitsas, J. Passerat-Palmbach, W. Bai, et al., DeepCut: Object segmentation from bounding box annotations using convolutional neural networks, IEEE Transactions on Medical Imaging 36 (2) (2017) 674–683.

[9] D. Lin, J. Dai, J. Jia, K. He, J. Sun, Scribblesup: Scribble-supervised convolutional networks for semantic segmentation, in: IEEE Conference on Computer Vision and Pattern Recognition (CVPR), 2016, pp. 3159–3167.

[10] H. Kervadec, J. Dolz, M. Tang, E. Granger, Y. Boykov, I. Ben Ayed, Constrained-CNN losses for weakly supervised segmentation, Medical Image Analysis 54 (2019) 88–99.

[11] D. Pathak, P. Krahenbuhl, T. Darrell, Constrained convolutional neural networks for weakly supervised segmentation, in: IEEE International Conference on Computer Vision (ICCV), 2015, pp. 1796–1804.

[12] M. Carbonneau, V. Cheplygina, E. Granger, G. Gagnon, Multiple instance learning: A survey of problem characteristics and applications, Pattern Recognition 77 (2018) 329–353.

[13] A.L. Bearman, O. Russakovsky, V. Ferrari, F. Li, What's the point: Semantic segmentation with point supervision, in: European Conference on Computer Vision (ECCV), Part VII, 2016, pp. 549–565.

[14] G. Papandreou, L. Chen, K.P. Murphy, A.L. Yuille, Weakly- and semi-supervised learning of a deep convolutional network for semantic image segmentation, in: IEEE International Conference on Computer Vision (ICCV), 2015, pp. 1742–1750.

[15] A. Khoreva, R. Benenson, J.H. Hosang, M. Hein, B. Schiele, Simple does it: Weakly supervised instance and semantic segmentation, in: IEEE Conference on Computer Vision and Pattern Recognition (CVPR), 2017, pp. 1665–1674.

[16] J. Weston, F. Ratle, H. Mobahi, R. Collobert, Deep learning via semi-supervised embedding, in: Neural Networks: Tricks of the Trade, Springer, 2012, pp. 639–655.

[17] D. Marin, M. Tang, I. Ben Ayed, Y. Boykov, Beyond gradient descent for regularized segmentation losses, in: IEEE conference on Computer Vision and Pattern Recognition (CVPR), 2019, pp. 10187–10196.

[18] Y. Zhou, Z. Li, S. Bai, X. Chen, M. Han, C. Wang, et al., Prior-aware neural network for partially-supervised multi-organ segmentation, in: IEEE International Conference on Computer Vision (ICCV), 2019, pp. 10671–10680.

[19] Z. Jia, X. Huang, E.I. Chang, Y. Xu, Constrained deep weak supervision for histopathology image segmentation, IEEE Transactions on Medical Imaging 36 (11) (2017) 2376–2388.

[20] P. Krähenbühl, V. Koltun, Efficient inference in fully connected CRFs with Gaussian edge potentials, in: Advances in Neural Information Processing Systems (NIPS), 2011, pp. 109–117.

[21] Y. Boykov, M.P. Jolly, Interactive graph cuts for optimal boundary and region segmentation of objects in n-d images, in: IEEE International Conference on Computer Vision (ICCV), 2001, pp. 105–112.

[22] C. Rother, V. Kolmogorov, A. Blake, GrabCut: Interactive foreground extraction using iterated graph cuts, ACM Transactions on Graphics 23 (3) (2004) 309–314.

[23] P. Baqué, T.M. Bagautdinov, F. Fleuret, P. Fua, Principled parallel mean-field inference for discrete random fields, in: IEEE Conference on Computer Vision and Pattern Recognition (CVPR), 2016, pp. 5848–5857.

[24] A. Adams, J. Baek, M.A. Davis, Fast high-dimensional filtering using the permutohedral lattice, Computer Graphics Forum 29 (2) (2010) 753–762.

[25] D. Marin, M. Tang, I. Ben Ayed, Y. Boykov, Kernel clustering: Density biases and solutions, IEEE Transactions on Pattern Analysis and Machine Intelligence 41 (1) (2018) 136–147.

[26] M. Tang, D. Marin, I. Ben Ayed, D. Marin, Y. Boykov, Kernel cuts: Kernel & spectral clustering meet regularization, International Journal of Computer Vision 127 (5) (2019) 477–511.

[27] J. Shi, J. Malik, Normalized cuts and image segmentation, IEEE Transactions on Pattern Analysis and Machine Intelligence 22 (8) (2000) 888–905.

[28] U. Von Luxburg, A tutorial on spectral clustering, Statistics and Computing 17 (4) (2007) 395–416.

[29] M. Tang, D. Marin, I. Ben Ayed, Y. Boykov, Normalized cut meets MRF, in: European Conference on Computer Vision (ECCV), Part II, 2016, pp. 748–765.

[30] M. Everingham, S.M.A. Eslami, L.J.V. Gool, C.K.I. Williams, J.M. Winn, A. Zisserman, The Pascal visual object classes challenge: A retrospective, International Journal of Computer Vision 111 (1) (2015) 98–136.

[31] L. Chen, G. Papandreou, I. Kokkinos, K. Murphy, A.L. Yuille, DeepLab: Semantic image segmentation with deep convolutional nets, atrous convolution, and fully connected CRFs, IEEE Transactions on Pattern Analysis and Machine Intelligence 40 (4) (2018) 834–848.

[32] Y. Boykov, O. Veksler, R. Zabih, Fast approximate energy minimization via graph cuts, IEEE Transactions on Pattern Analysis and Machine Intelligence 23 (11) (2001) 1222–1239.

[33] O. Veksler, Efficient graph cut optimization for full CRFs with quantized edges, IEEE Transactions on Pattern Analysis and Machine Intelligence 42 (4) (2020) 1005–1012.

[34] A. Kolesnikov, C.H. Lampert, Seed, expand and constrain: Three principles for weakly-supervised image segmentation, in: European Conference on Computer Vision (ECCV), Part IV, 2016, pp. 695–711.

[35] L.C. Chen, G. Papandreou, I. Kokkinos, K. Murphy, A.L. Yuille, Semantic image segmentation with deep convolutional nets and fully connected CRFs, in: International Conference on Learning Representations (ICLR), 2015, pp. 1–14.

[36] S. Boyd, N. Parikh, E. Chu, B. Peleato, J. Eckstein, Distributed optimization and statistical learning via the alternating direction method of multipliers, Foundations and Trends in Machine Learning 3 (1) (2011) 1–122.

[37] P. Krähenbühl, V. Koltun, Parameter learning and convergent inference for dense random fields, in: International Conference on Machine Learning (ICML), 2013, pp. 513–521.

CHAPTER 10

Constrained deep networks

10.1 Weakly supervised segmentation via constrained CNNs

In the previous chapter, we discussed how conditional random field losses can be very helpful in the context of weakly supervised CNN segmentation [1–4]. More generally, imposing prior knowledge on the probability outputs of a deep network can mitigate the lack/uncertainty of annotations, and is a well-established general principle in machine learning [5,6]. Acting as regularizers, such priors leverage unlabeled samples with domain-specific knowledge. For instance, in medical imaging, it is common to have some prior knowledge about the size (or volume) of the target organ. Such a knowledge does not have to be precise, and may take the imprecise form of lower and upper bounds on segmentation-region size [7–9]. As we will see in the experiments discussed at the end of this chapter, enforcing inequality constraints on the target-region size, which can be expressed as a function of the network outputs, is very useful in the context of weakly supervised CNN segmentation.

In this chapter, we discuss weakly supervised semantic segmentation problems of the following general form, in which we optimize a partial-supervision loss subject to a set of N inequality constraints on the network outputs:

$$\min_{\boldsymbol{\theta}} \quad \sum_{p \in \Omega_{\mathcal{L}}} \mathcal{H}(\mathbf{z}_p, \mathbf{s}_p)$$
$$\text{s.t.} \quad f_i(\mathbf{s}) \leq 0, \ i = 1, \ldots N, \tag{10.1}$$

where we recall the following notations, which we used in the previous chapter:

- Vector $\mathbf{s}_p = (s_{p,0}, \ldots, s_{p,L-1})^t \in [0,1]^L$ contains the softmax outputs of the network for pixel p. Each $s_{p,l}$ denotes a probability of assigning a label $l \in \{0, \ldots, L-1\}$ to a pixel p:

$$s_{p,l} = \frac{1}{R_p} \exp r_p(l; \boldsymbol{\theta}, \mathbf{f}),$$

with $\boldsymbol{\theta}$ a vector containing the network trainable parameters, $\mathbf{f} = (f_p)_{p \in \Omega}$ the input image defined over spatial domain $\Omega \subset \mathbb{R}^{2,3}$, r_p the network output for pixel p and R_p a normalization constant. For notation simplicity, we will omit the dependence of s_p on $\boldsymbol{\theta}$ and $\mathbf{f}$ because this does not result in any ambiguity in subsequent discussions. Vector $\mathbf{s} = (\mathbf{s}_p)_{p \in \Omega} \in [0, 1]^{L|\Omega|}$ concatenates the network predictions for all the pixels in Ω.

- $\mathbf{z}_p = (z_{p,0}, \dots, z_{p,L-1})^t \in \{0, 1\}^L$ is a binary vector indicating the ground-truth of a labeled point p, with $z_{p,l} = 1$ if pixel p has label l and $z_{p,l} = 0$ otherwise.

- Similarly to the previous chapter, $\mathcal{H}$ evaluates some divergence between the output probabilities of the network and the ground-truth for the portion of pixels that have labels. A standard choice is the cross–entropy:

$$\mathcal{H}(\mathbf{z}_p, \mathbf{s}_p) = -\sum_l z_{p,l} \ln s_{p,l}.$$

However, the subsequent discussions of constrained CNNs are not limited to a specific form of $\mathcal{H}$. $\Omega_{\mathcal{L}} \subset \Omega$ denotes the set of labeled pixels, which corresponds to a fraction of the spatial domain Ω of a given training image. The difference with full supervision is that the summation in the objective in (10.1) is over the subset of pixels that are labeled, $\Omega_{\mathcal{L}}$, not over whole image domain Ω. For instance, such partial labels can be obtained with few scribbles; see the examples in Chapter 9.

Inequality constraints of the general form in (10.1) can embed very useful priors on the network predictions for unlabeled points via functions f_i. Such priors could come from some domain-specific knowledge about the target regions. Assume, for instance, that we have prior knowledge about the size (or cardinality) of the target segmentation region (or class) l. Such a knowledge does not have to be precise; it can be in the form of lower and upper bounds on size, which is common in medical-image segmentation problems [8,9]. For instance, when we have an upper bound a on the size of region l, we can impose the following constraint:

$$\sum_{p \in \Omega} s_{p,l} - a \le 0.$$

In this case, the corresponding constraint i in the general-form in Eq. (10.1) uses the following function:

$$f_i(\mathbf{s}) = \sum_{p \in \Omega} s_{p,l} - a.$$

In a similar way, one can impose a lower bound b on the size of region l using the following function:

$$f_i(\mathbf{s}) = b - \sum_{p \in \Omega} s_{p,l}.$$

The same form of constraints can impose image-tag priors, a from of weak supervision enforcing whether a target region is present or absent in a given training image, as in multiple instance learning (MIL) scenarios [7,10, 11], which we discussed in the previous chapter. For instance, a suppression constraint, which encourages that region l does not appear in the image, corresponds to using the following constraint function f_i in model (10.1):

$$f_i(\mathbf{s}) = \sum_{p \in \Omega} s_{p,l}.$$

The presence of region l in an image could be enforced via the following constraint function f_i:

$$f_i(\mathbf{s}) = 1 - \sum_{p \in \Omega} s_{p,l}.$$

10.2 Constraint optimization

Even when the constraints are convex with respect to the probability outputs of the network, the problem in Eq. (10.1) is challenging for deep CNNs that are common in modern computer vision problems, which involve millions of parameters, as is the case in semantic segmentation. In the general context of optimization, a standard technique to deal with hard inequality constraints is to solve the Lagrangian primal and dual problems in an alternating scheme [12]. For problem (10.1), this amounts to alternating the optimization of a CNN for the primal with stochastic optimization, e.g., SGD, and projected gradient-ascent iterates for the dual. However, despite the clear benefits of imposing constraints on CNNs, such a standard Lagrangian-dual optimization is avoided in the context of modern deep networks due, in part, to computational-tractability issues. As pointed out in [10,13], there is a consensus within the community that imposing hard constraints on the outputs of deep CNNs that are common in modern computer vision problems is impractical: The use of Lagrangian-dual optimization for networks with millions of parameters requires training a whole CNN after each iterative dual step [10]. In fact, Lagrangian methods were investigated for constrained neural networks a long before the deep learning age [14,15]. Based on solving a large linear system of equations, these early methods are not applicable to deep CNNs, in which case they would have to deal with very large matrices [13].

Tackling hard inequality or equality constraints with Lagrangian optimization for modern deep CNNs is still in a nascent stage, and only a few recent works focused on the subject [10,13,16]. For instance, the work in [13] imposed hard equality constraints on 3D human–pose estimation. To tackle the computational difficulty, the authors used a Kyrlov subspace approach and limited the solver to only a randomly selected subset of the constraints within each iteration. Therefore, constraints that are satisfied at one iteration may not be satisfied at the next. The work of Pathak et al. [10] tackled dual optimization efficiently in the context of weakly supervised CNN segmentation. The authors proposed an approximate solution, which imposes inequality constraints on a latent distribution rather than directly on the network probability outputs. This latent distribution yields synthesized ground-truth labels, similar to the segmentation proposals discussed in the previous chapter in the context of CRFs. Then, a *single* CNN is trained so as to minimize the KL divergence between the network probability outputs and the latent distribution, mimicking full supervision and easing the computational burden of Lagrangian optimization. The training of the CNN alternates two substeps: (i) generating the proposals under the constraints, with the network parameters fixed[1]; and (ii) standard SGD substeps for learning the parameters of the networks with the proposals, i.e., synthesized ground-truth labels, fixed. We will discuss the work of Pathak et al. [10] in greater detail later in this chapter.

In the context of deep networks, equality or inequality constraints are typically handled in a "soft" manner by augmenting the loss with a *penalty* function [7,17,18]. The penalty-based approach is a simple alternative to Lagrangian optimization and is well known in the general context of constrained optimization; see [19], Chapter 4. In general, such penalty-based methods approximate a constrained minimization problem with an unconstrained one by adding a term, which increases when the constraints are violated. This is convenient for deep networks because it removes the requirement for explicit Lagrangian-dual optimization. The inequality constraints are fully handled within stochastic optimization, as in standard unconstrained losses, avoiding gradient ascent iterates/projections over the

---

[1] Note that, when the constraints are convex, this subproblem is convex.

dual variables and reducing the computational load for training. However, this simplicity of penalty methods comes at a price. In fact, it is well known that penalty methods are not optimal: They do not guarantee constraint satisfaction and require careful and *ad hoc* tuning of the relative importance (or weight) of the penalty term in the overall function that is being minimized. Lagrangian optimization can deal with these difficulties and has several well-known theoretical and practical advantages over penalty-based methods [20,21]: It finds automatically the optimal weights of the constraints and guarantees constraint satisfaction when feasible solutions exist. Unfortunately, as pointed out recently in [7,13], these advantages of Lagrangian optimization do not materialize in practice in the context of deep networks. Apart from the computational-feasibility aspects, which the recent works in [10,13] address to some extent with approximations, the performances of Lagrangian optimization are, surprisingly, below those obtained with simple, much less computationally intensive penalties [7,13]. This is, for instance, the case of the weakly supervised CNN semantic segmentation results in [7], which showed that a simple quadratic-penalty formulation of inequality constraints outperforms substantially the Lagrangian method in [10]. We will discuss this penalty-based approach and its experiments in greater detail later in this chapter. Also, the authors of [13] reported surprising results in the context of 3D human pose estimation. In their case, replacing the equality constraints with simple quadratic penalties yielded better results than Lagrangian optimization.

In the context of constrained deep networks, and as hypothesized recently in [13], the difficulty in Lagrangian optimization might be due, in part, to the interplay between stochastic optimization for the primal problem and the iterates/projections for the dual. The latter handles the constraints via basic (nonstochastic) projected gradient-ascent methods. Basic gradient methods have well-known issues with deep networks, e.g., they are sensitive to the learning rate and prone to weak local minima. Therefore, the dual part in Lagrangian optimization might obstruct the practical and theoretical benefits of stochastic optimization (e.g., speed and strong generalization performance), which are widely established for unconstrained deep network losses [22]. Penalty methods deal with the constraints fully within stochastic optimization, avoiding the Lagrangian-dual iterates and projections. This might explain why simple penalty-based methods, despite their well-known limitations in optimization, outperform Lagrangian approaches in the context constrained deep CNNs [7,13].

10.2.1 Penalty-based methods

Penalty-based methods approximate a constrained problem of the form (10.1) with an unconstrained one, which can be tackled with standard optimization techniques such as stochastic gradient descent for deep networks:

$$\min_{\theta} \sum_{p \in \Omega_L} \mathcal{H}(\mathbf{z}_p, \mathbf{s}_p) + \gamma \mathcal{P}(\mathbf{s}), \tag{10.2}$$

where γ is a positive constant and $\mathcal{P}$ is a penalty function satisfying the following conditions:

- $\mathcal{P}$ is positive, continuous and differentiable; and
- $\mathcal{P}(\mathbf{s}) = 0$ if and only if $\mathbf{s}$ satisfies all the constraints.

Function $\mathcal{P}$ results in an increase of augmented loss (10.2) when a constraint is not satisfied, i.e., minimizing (10.2) encourages satisfaction of the inequality constraints of the original problem in (10.1).

It is common to use a quadratic penalty, which takes the following form for inequality constraints [7,17]:

$$\mathcal{P}(\mathbf{s}) = \sum_{1}^{N} [f_i(\mathbf{s})]_+^2, \tag{10.3}$$

where $[x]_+ = \max(0, x)$ denotes the rectifier function. Fig. 10.1 depicts an illustration of this quadratic penalty function for one constraint. In this case, an unsatisfied constraint increases loss (10.2) by the square of constraint violation, multiplied by weight γ. Of course, although commonly used, this quadratic penalty is not a unique choice, and there are other options [19].

The difficulty with a basic penalty approach is that one has to choose parameter γ in Eq. (10.2) in an *ad hoc* way. Obviously, the effect of the penalty term is counter-balanced by the partial cross-entropy. Therefore, on the one hand, choosing a very small γ is not likely to yield a solution that satisfies the constraints of the original problem in (10.1). On the other hand, a very large γ will certainly find feasible solutions but ignores learning from the partial labels, which may not yield satisfying results.

A differentiable penalty $\mathcal{P}(\mathbf{s})$ accommodates standard stochastic gradient descent. Let us examine the effect of a quadratic penalty during gradient-descent steps for a simple constraint in the form of an upper bound on the size of the target region. In this case, the penalty takes the following form:

$$\mathcal{P}(\mathbf{s}) \propto \left[\sum_{p \in \Omega} s_{p,l} - a \leq 0 \right]_+^2.$$

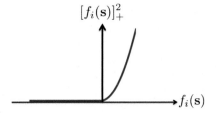

Figure 10.1 Illustration of the quadratic penalty for inequality-constrained optimization.

The corresponding gradient-descent update uses the following derivative:

$$-\frac{\partial \mathcal{P}(\mathbf{s})}{\partial \boldsymbol{\theta}} \propto \begin{cases} \left(a - \sum_{p \in \Omega} s_{p,l}\right) \sum_{p \in \Omega} \frac{\partial s_{p,l}}{\partial \boldsymbol{\theta}}, & \text{if } \sum_{p \in \Omega} s_{p,l} > a, \\ 0, & \text{otherwise,} \end{cases} \quad (10.4)$$

where $\frac{\partial s_{p,l}}{\partial \boldsymbol{\theta}}$ is the standard derivative of network probability output $s_{p,l}$. The expression of the gradient in (10.4) has a straightforward interpretation. Notice that, when the size constraint is satisfied during gradient-descent steps, i.e., $\sum_{p \in \Omega} s_p \leq a$, we have $\frac{\partial \mathcal{P}(\mathbf{s})}{\partial \boldsymbol{\theta}} = 0$. In this case, the gradient ensuing from the penalty does not have any effect on the current update of the parameters of the deep network. In the case of constraint violation, i.e., when the size of the current target region exceeds its upper bound, the current set of network parameters $\boldsymbol{\theta}$ verifies $\sum_{p \in \Omega} s_{p,l} > a$. In this case, quantity $\left(a - \sum_{p \in \Omega} s_{p,l}\right)$ is negative, and the first line of (10.4) yields a gradient descent update on the sum of probability outputs $s_{p,l}$, encouraging each of these outputs to decrease. This makes sense because it encourages the size of the target region to decrease, until it reaches constraint satisfaction.

10.2.2 Lagrangian optimization via proposals

Pathak et al. [10] tackled Lagrangian optimization efficiently in the context of weakly supervised CNN segmentation. The method can be viewed as an approximate solution to constrained problem (10.1) in which the inequality constraints are imposed on a latent variable rather than directly on the network probability outputs. This latent variable, which we denote $\mathbf{u} \in [0, 1]^{L |\Omega|}$ in what follows, describes synthesized ground-truth labels (or segmentation proposals), in a way similar to the alternating direction method (ADM) interpretation discussed the previous chapter in the context of conditional random fields (CRFs). Such a latent variable takes the same form as in Chapter 9: $\mathbf{u} = (\mathbf{u}_p)_{p\in\Omega}$, with each vector $\mathbf{u}_p = (u_{p,0}, \ldots, u_{p,L-1})^t \in [0,1]^L$ being within L-dimensional probability simplex $\nabla_L = \{\mathbf{y} \in [0,1]^L \mid \mathbf{1}'\mathbf{y} = 1\}$. The method trains a single CNN by minimizing the KL divergence between the network probability outputs and the latent variable, subject to the constraints:

$$\min_{\boldsymbol{\theta},\mathbf{u}} \quad \sum_{p\in\Omega} \mathrm{KL}(\mathbf{u}_p\|\mathbf{s}_p)$$

$$\text{s.t.} \quad f_i(\mathbf{u}) \leq 0, \ i = 1, \ldots N. \tag{10.5}$$

Notice that, unlike in the original problem (10.1), the constraints here are imposed on latent variable $\mathbf{u}$, not on the network outputs. The key idea here is that minimizing the KL divergence encourages the output of the deep network to match latent distribution $\mathbf{u}$ as closely as possible. Therefore, one can impose the constraints on $\mathbf{u}$ instead of $\mathbf{s}$, while training a CNN in a way that resembles full supervision. The training of the CNN alternates two substeps: (1) fixing the network parameters and generating a segmentation proposal (synthesized ground truth) by optimizing (10.5) w.r.t. $\mathbf{u}$, which is a convex problem if the constraints are convex; and (2) standard stochastic gradient descent substeps for optimizing with respect to network parameters $\boldsymbol{\theta}$, with the proposal fixed.

Let us consider the constrained problem (10.5) in the case of linear constraints taking the general form:

$$f_i(\mathbf{u}) = \sum_{p\in\Omega} \mathbf{u}_p^t \mathbf{g}_p - a,$$

where $\mathbf{g}_p = (g_{p,0}, \ldots, g_{p,L-1})^t \in \mathbb{R}^L$ denotes some given unary potentials. For example, the size constraint we discussed earlier is a particular case of this general linear form. To impose an upper bound a on the size of target region l, one can take $g_{p,l} = 1$ and $g_{p,k} = 0$, for $k \neq l$. With the network parameters fixed, optimizing (10.5) subject to linear inequality and simplex constraint becomes:

$$\min_{\mathbf{u}} \quad \sum_{p\in\Omega} \left[-\mathbf{u}_p^t \ln(\mathbf{s}_p) + \mathbf{u}_p^t \ln(\mathbf{u}_p) \right]$$

$$\text{s.t.} \quad \sum_{p\in\Omega} \mathbf{u}_p^t \mathbf{g}_p - a \leq 0; \ \mathbf{u}_p \in \nabla_L \ \forall p = 1, \ldots, |\Omega|, \tag{10.6}$$

where, we recall, $\nabla_L = \{\mathbf{y} \in [0,1]^L \mid \mathbf{1}'\mathbf{y} = 1\}$ denotes the L-dimensional probability simplex. Notice that, in the sum of the objective in (10.6), each negative entropy term $\mathbf{u}_p^t \ln(\mathbf{u}_p)$ acts as a barrier function (or a penalty) restricting the domain of $\mathbf{u}_p$ to nonnegative values. This avoids imposing explicitly simplex constraints $\mathbf{u}_p \geq 0$. Ignoring these constraints, we can write the Lagrangian dual corresponding to (10.6) as follows:

$$\mathcal{L}(\mathbf{u}, \boldsymbol{\alpha}, \beta) = \sum_{p \in \Omega} \left[-\mathbf{u}_p^t \ln(\mathbf{s}_p) + \mathbf{u}_p^t \ln(\mathbf{u}_p) + \alpha_p (\mathbf{1}^t \mathbf{u}_p - 1) \right] + \beta \left(\sum_{p \in \Omega} \mathbf{u}_p^t \mathbf{g}_p - a \right).$$
(10.7)

$|\Omega|$-dimensional vector $\boldsymbol{\alpha} = (\alpha_p)_{p \in \Omega}$ contains the Lagrange multipliers for the simplex constraints, and β is the multiplier for the inequality constraint. Here, we limit the discussion to a single inequality constraint for the sake of clarity. However, extending the formulation to multiple constraints is straightforward. The objective in (10.6) is convex, and the constraints are linear. Therefore, strong duality holds if feasible points exist. In this case, the solution of the primal problem is the same as the dual one:

$$\min_{\mathbf{u}} \max_{\beta \geq 0, \boldsymbol{\alpha}} \mathcal{L}(\mathbf{u}, \boldsymbol{\alpha}, \beta) = \max_{\beta \geq 0, \boldsymbol{\alpha}} \min_{\mathbf{u}} \mathcal{L}(\mathbf{u}, \boldsymbol{\alpha}, \beta).$$

To compute the dual function, we optimize the Lagrangian with respect to $\mathbf{u}$:

$$\min_{\mathbf{u}} \mathcal{L}(\mathbf{u}, \boldsymbol{\alpha}, \beta).$$

As $\mathcal{L}$ is convex w.r.t. $\mathbf{u}$, the global optimum can be obtained by setting the derivative equal to zero. Notice that $\mathcal{L}$ is separable over variables $\mathbf{u}_p$. In fact, we can write $\mathcal{L}$, up to a constant w.r.t. $\mathbf{u}$, in the form of a sum of independent functions, each corresponding to one pixel p:

$$-\mathbf{u}_p^t \ln \mathbf{s}_p + \mathbf{u}_p^t \ln \mathbf{u}_p + \beta \mathbf{u}_p^t \mathbf{g}_p + \alpha_p \mathbf{1}^t \mathbf{u}_p.$$
(10.8)

Setting the derivative of (10.8) w.r.t. $\mathbf{u}_p$ equal to zero gives a closed-form solution:

$$u_{p,l}^* = s_{p,l} \exp^{-\beta g_{p,l} - \alpha_p - 1}.$$
(10.9)

This yields the dual function for constrained problem (10.6):

$$\mathcal{D}(\boldsymbol{\alpha}, \beta) = \min_{\mathbf{u}} \mathcal{L}(\mathbf{u}, \boldsymbol{\alpha}, \beta) = \mathcal{L}(\mathbf{u}^*, \boldsymbol{\alpha}, \beta),$$
(10.10)

where $\mathbf{u}^* = (\mathbf{u}_p^*)_{p \in \Omega}$, and $\mathbf{u}_p^* = (u_{p,0}^*, \ldots, u_{p,L-1}^*)^t$. The dual function is concave w.r.t. dual variables $\boldsymbol{\alpha}, \beta$ (minimum of linear functions). Maximizing the dual function w.r.t. $\boldsymbol{\alpha}$ can be done in closed-form by setting the derivative of $\mathcal{D}(\boldsymbol{\alpha}, \beta)$ w.r.t. each α_p equal to zero, which yields simplex constraints $\mathbf{1}^t\mathbf{u}_p^* - 1 = 0 \forall p$. Plugging expression (10.9) into the simplex constraint yields, after some manipulations, the following optimality condition over each α_p: $\exp^{-\alpha_p-1} = 1/\sum_l \mathbf{s}_{p,l} \exp^{-\beta g_{p,l}}$. Substituting this expression into (10.9) yields the following solution:

$$u_{p,l}^* = \frac{s_{p,l} \exp^{-\beta g_{p,l}}}{\sum_{l'} s_{p,l'} \exp^{-\beta g_{p,l'}}}. \tag{10.11}$$

Now, with this expression of the optimal segmentation proposals, the dual function depends only on β: $\mathcal{D}(\beta) = \mathcal{L}(\mathbf{u}^*, \beta)$. As the dual function is concave w.r.t. β, it can be optimized globally with projected gradient ascent. The gradient of the dual function w.r.t. to β is:

$$\nabla \mathcal{D}(\beta) = \sum_{p \in \Omega} (\mathbf{u}_p^*)^t \mathbf{g}_p - a. \tag{10.12}$$

The projected gradient ascent solves $\max_{\beta \geq 0} \mathcal{D}(\beta)$ by the following updates:

$$\beta^{i+1} = \max(0, \beta^i + \rho \nabla \mathcal{D}(\beta^i)), \tag{10.13}$$

where i denotes the iteration counter and ρ is a nonnegative step size. Finally, optimizing (10.5) is done by training a CNN, which alternates between two substeps: (i) generating a segmentation proposal $\mathbf{u}$ under the constraints, with network parameters $\boldsymbol{\theta}$ fixed; and (ii) standard stochastic gradient descent substeps for learning parameters $\boldsymbol{\theta}$ with the segmentation proposal $\mathbf{u}$ fixed. The constrained-CNN training process is summarized in the following:

1. Initialize β to 0;
2. Initialize $\mathbf{u}$ (e.g., using image-level tags);
3. Iterate the following steps:
   a. Perform standard stochastic gradient descent substeps for updating $\boldsymbol{\theta}$, with $\mathbf{u}$ fixed. This becomes similar to optimizing a standard cross-entropy loss with the ground-truth labels given by segmentation proposal $\mathbf{u}$, as in supervised learning:

$$\arg\min_{\boldsymbol{\theta}} \sum_{p \in \Omega} \mathcal{H}(\mathbf{u}_p, \mathbf{s}_p);$$

   b. Update $\mathbf{u}$ by iterating the following two steps until convergence:

- Solve $\mathbf{u}^*$ analytically with expressions (10.11) and
- Update β with projected gradient ascent (10.13).

The update expressions in (10.11), (10.12), and (10.13) have a straightforward interpretation in the case of a constraint of the form $\sum_{p \in \Omega} \mathbf{u}_p^t \mathbf{g}_p - a \leq 0$. Without a loss of generality, let us assume that $\mathbf{g}_p$ is nonnegative, as is the case of an upper-bound constraint on the target region size. Notice that the gradient of $\mathcal{D}$ is equal to the current value of the constraint, i.e., $\sum_{p \in \Omega} \mathbf{u}_p^t \mathbf{g}_p - a$. Therefore, in the case of constraint violation, i.e., $\sum_{p \in \Omega} \mathbf{u}_p^t \mathbf{g}_p - a > 0$, the current value of this gradient is positive. In this case, the gradient-ascent update in Eq. (10.13) increases the value of Lagrange multiplier β. This contributes to decreasing segmentation-proposal variables according to Eq. (10.11). Clearly, this decreases the value of $\sum_{p \in \Omega} \mathbf{u}_p^t \mathbf{g}_p - a$, encouraging constraint satisfaction.

10.3 Discussion of some experimental results

In this section, we discuss the experiments in [7] that focused on left ventricular endocardium segmentation in MRI. The experiments are based on a publicly available data set, which contains 100 MRI exams: 75 used for training the algorithms and 25 for validation. To evaluate constrained CNNs in the weakly supervised setting, partial labels are generated synthetically and correspond to only 0.1% of the labels in the fully supervised setting. Fig. 10.2 shows examples of various levels of supervision, with fully labeled images at the top and the corresponding partial (weak) labels at the bottom. In the fully supervised setting, all the pixels are labeled, with green indicating the target region (foreground) and red indicating the background. In the weakly supervised setting, we know the labels of only a fraction of the foreground pixels, which are depicted in green color. Several models are trained with different levels of supervision and constraints, using both Lagrangian proposals and quadratic penalties. First, as a baseline, a network is trained with the partial cross-entropy objective (CE) in (10.1), using only the fraction of labeled pixels without any constraint. Then, with the same partially labeled images, image-tag constraints can enforce whether a target region is present or absent in a given training image, as in MIL scenarios. As discussed earlier, a suppression constraint, which takes the form of $f_i(\mathbf{s}) = \sum_{p \in \Omega} s_{p,l}$, encourages that region l does not appear in the image, whereas $f_i(\mathbf{s}) = 1 - \sum_{p \in \Omega} s_{p,l}$ encourages the presence of the region. Finally, the weak-supervision experiments included evaluations of constraints in the form of lower and upper bounds on the size of the target

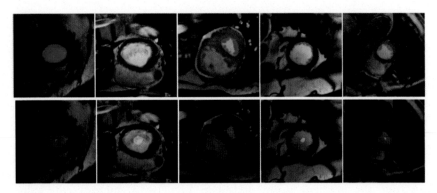

Figure 10.2 Different supervision levels. Top: fully labeled images, with green indicating the target endocardium region (foreground) and red indicating the background; Bottom: weakly supervised learning, with only a fraction of the foreground pixels having known labels, which are depicted in green color.

region. Recall that, in the case of an upper bound on size, we can impose inequality constraint $\sum_{p \in \Omega} s_{p,l} - a \leq 0$, whereas, in the case of a lower bound, we can use $b - \sum_{p \in \Omega} s_{p,l} \leq 0$. Such a prior knowledge on region size is learned from the fully labeled data of a single subject. The weak-supervision results are juxtaposed to the full-supervision performance, i.e., to the segmentation results obtained when the labels of all pixels (both the endocardium and background) are known during training, with full cross-entropy being optimized. The evaluation is based on the commonly used Dice similarity coefficient (DSC),[2] which assesses a similarity between an obtained segmentation region $S \subset \Omega$ and a ground-truth segmentation region $G \subset \Omega$: $\mathrm{DSC} = \frac{2|S \cap G|}{|S| + |G|}$, where $|.|$ denotes cardinality. The higher the DSC, the better the result. The measure takes its values in $[0, 1]$, with a value of 1 indicating a perfect match with the ground truth and a value of 0 indicating a total mismatch.

Table 10.1 reports the results (mean DSC) of [7] on the validation set for all the weakly-supervised constrained-CNN models, juxtaposing the Lagrangian proposals and the quadratic penalty (direct loss) for handling the constraints. Also, Fig. 10.3 plots the evolution of the mean DSC during training using validation samples and different constrained-CNN models. Using the partial cross-entropy based on a fraction of labeled endocardium pixels yielded, as expected, a very low mean DSC. The use of image-

---

[2] DSC is a standard measure for evaluating medical-image segmentation algorithms.

Table 10.1 Cardiac image segmentation results (mean DSC on the validation set) with different levels of supervision and different techniques for imposing constraints, i.e., Lagrangian proposals and quadratic penalties.

	Model	Method	DSC
	Partial CE		0.07
	CE + Tags	Lagrangian proposals [10]	0.61
Weakly supervised	Partial CE + Tags	Direct loss [7]	0.71
	CE + Tags + Size*	Lagrangian proposals [10]	0.61
	Partial CE + Tags + Size*	Direct loss [7]	0.81
	CE + Tags + Size**	Lagrangian proposals [10]	0.65
	Partial CE + Tags + Size**	Direct loss [7]	0.84
Fully supervised	Cross–entropy		0.92

*Upper bound / **Lower and upper bounds. CE refers to cross-entropy.

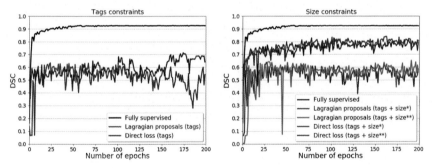

Figure 10.3 Evolution of the mean DSC during training, using validation samples and different constrained-CNN models.

tag (MIL) constraints increased substantially the performance, reaching a mean DSC of 0.71 for the quadratic penalty. By adding constraints on the size of the target region, one can, interestingly, reach over 90% of the full-supervision performance with only 0.1% of the pixels labeled. Surprisingly, the levels of performance of Lagrangian proposals are substantially lower than those obtained with the simple, less computationally intensive quadratic penalties, although such penalties do not guarantee constraint satisfaction. For instance, in the case of image-tag constraints, the penalty approach increased the mean DSC by 15% in comparison to Lagrangian proposals. The difference between the two approaches becomes more im-

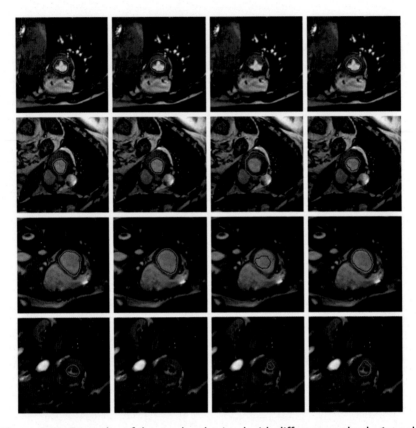

Figure 10.4 Examples of the results obtained with different methods. 1st column: the ground-truth segmentations; 2nd column: fully supervised learning; 3rd column: weakly supervised learning (Lagrangian proposals); 4th column: weakly supervised learning (quadratic penalties). (*The figures are taken from [7].*)

portant (30%) when adding region–size constraints. Also, there are other observations that one can make from Fig. 10.3. Typically, Lagrangian optimization reaches its best performance faster than the penalty approach. More importantly, one can also observe that the evolution of the mean DSC during training for the Lagrangian approach is less stable than with the penalty method. Fig. 10.4 depicts a sample of the results for visual inspection. For these examples, the weak-supervision results are based on partial cross-entropy with several constraints, including image tags and lower/upper bounds on the size of the target region.

References

[1] D. Marin, M. Tang, I. Ben Ayed, Y. Boykov, Beyond gradient descent for regularized segmentation losses, in: IEEE conference on Computer Vision and Pattern Recognition (CVPR), 2019, pp. 10187–10196.

[2] M. Tang, F. Perazzi, A. Djelouah, I. Ben Ayed, C. Schroers, Y. Boykov, On regularized losses for weakly-supervised CNN segmentation, in: European Conference on Computer Vision (ECCV), Part XVI, 2018, pp. 524–540.

[3] M. Rajchl, M.C.H. Lee, O. Oktay, K. Kamnitsas, J. Passerat-Palmbach, W. Bai, et al., DeepCut: Object segmentation from bounding box annotations using convolutional neural networks, IEEE Transactions on Medical Imaging 36 (2) (2017) 674–683.

[4] D. Lin, J. Dai, J. Jia, K. He, J. Sun, Scribblesup: Scribble-supervised convolutional networks for semantic segmentation, in: IEEE Conference on Computer Vision and Pattern Recognition (CVPR), 2016, pp. 3159–3167.

[5] J. Weston, F. Ratle, H. Mobahi, R. Collobert, Deep learning via semi-supervised embedding, in: Neural Networks: Tricks of the Trade, Springer, 2012, pp. 639–655.

[6] I. Goodfellow, Y. Bengio, A. Courville, Deep Learning, MIT Press, 2016.

[7] H. Kervadec, J. Dolz, M. Tang, E. Granger, Y. Boykov, I. Ben Ayed, Constrained-CNN losses for weakly supervised segmentation, Medical Image Analysis 54 (2019) 88–99.

[8] M. Niethammer, C. Zach, Segmentation with area constraints, Medical Image Analysis 17 (1) (2013) 101–112.

[9] L. Gorelick, F.R. Schmidt, Y. Boykov, Fast trust region for segmentation, in: IEEE Conference on Computer Vision and Pattern Recognition (CVPR), 2013, pp. 1714–1721.

[10] D. Pathak, P. Krahenbuhl, T. Darrell, Constrained convolutional neural networks for weakly supervised segmentation, in: IEEE International Conference on Computer Vision (ICCV), 2015, pp. 1796–1804.

[11] G. Papandreou, L. Chen, K.P. Murphy, A.L. Yuille, Weakly- and semi-supervised learning of a deep convolutional network for semantic image segmentation, in: IEEE International Conference on Computer Vision (ICCV), 2015, pp. 1742–1750.

[12] S. Boyd, L. Vandenberghe, Convex Optimization, Cambridge University Press, 2004.

[13] P. Márquez-Neila, M. Salzmann, P. Fua, Imposing hard constraints on deep networks: Promises and limitations, in: CVPR Workshop on Negative Results in Computer Vision, 2017, pp. 1–9.

[14] S. Zhang, A. Constantinides, Lagrange programming neural networks, IEEE Transactions on Circuits and Systems II: Analog and Digital Signal Processing 39 (7) (1992) 441–452.

[15] J.C. Platt, A.H. Barr, Constrained differential optimization, Tech. rep., California Institute of Technology, 1988.

[16] Y. Nandwani, A. Pathak, Mausam, P. Singla, A primal dual formulation for deep learning with constraints, in: Neural Information Processing Systems (NeurIPS), 2019, pp. 12157–12168.

[17] F.S. He, Y. Liu, A.G. Schwing, J. Peng, Learning to play in a day: Faster deep reinforcement learning by optimality tightening, in: International Conference on Learning Representations (ICLR), 2017, pp. 1–13.

[18] Z. Jia, X. Huang, E.I. Chang, Y. Xu, Constrained deep weak supervision for histopathology image segmentation, IEEE Transactions on Medical Imaging 36 (11) (2017) 2376–2388.

[19] D.P. Bertsekas, Nonlinear Programming, Athena Scientific, Belmont, MA, 1995.

[20] R. Fletcher, Practical Methods of Optimization, John Wiley & Sons, 1987.

[21] P. Gill, W. Murray, M. Wright, Practical Optimization, Academic Press, 1981.

[22] M. Hardt, B. Recht, Y. Singer, Train faster, generalize better: Stability of stochastic gradient descent, in: International Conference on Machine Learning (ICML), 2016, pp. 1225–1234.

Index

Symbols

α-β swap, 24, 26
α-expansion moves, 24–26
α-expansion solver, 139, 140
   discrete, 134, 139, 140

A

Affinity matrix, 128, 131
   pairwise, 128
   PSD, 94
   symmetric, 94
Alternating direction method (ADM), 135,
   136, 151
   minimization, 138
   optimization, 138
   splitting, 138
Alternating optimization, 136
   schemes, 63
Alternating-direction method of multipliers
   (ADMM), 137, 138
Aorta segmentations, 78
Assignment variables, 17, 38, 136
Average association (AA), 78, 79, 92, 94

B

Background
   GMMs, 55
   likelihood, 6
   pixel, 2
   probability, 5, 54
   regions, 9
   segmentation, 54
Bags, 125
Balanced segmentation, 64
Binary
   hidden variable, 2
   indicator variables, 127
   optimization variable, 26
   pairwise functions, 18
   segmentation, 7, 11, 12, 19, 111
      scenario, 6
      variable, 6
   vector, 54, 69, 78, 85, 92, 95, 111, 124,
      135, 146

Bound
   optimization, 35, 85, 87, 89, 90, 96, 102,
      103, 107, 108
   optimizers, 35, 87–89, 91
   relaxation, 99
Boykov–Kolmogorov (BK) algorithm, 12,
   13, 22, 23
Breadth computer vision, 4, 86, 123

C

Cardinality, 25, 48, 67, 90, 100, 146
   prior, 100, 101, 105
   region, 56, 61, 100
Chan–Vese
   functional, 49
   model, 46, 47
   objective, 49
Classical computer vision, 132
Clique potential, 4
Colors
   bins, 56, 80, 104, 105
   features, 5, 56, 62
   histograms, 6, 91
   segmentation, 63
Combinatorial optimization, 19, 21
   algorithms, 12, 18
   techniques, 45
Computed tomography (CT), 73, 77
Computer vision, 1, 2, 4, 7, 11, 13, 22–24,
   33, 46, 67, 70, 71, 79, 86, 87, 94,
   111, 123, 124
   algorithms, 12, 18, 86
   applications, 1, 19, 22, 96
   breadth, 4, 86, 123
   classical, 132
   problems, 1, 2, 12, 23, 27, 38, 147
   tasks, 2, 13
   works, 87
Conditional entropy, 60–62, 65, 66
   features, 60
Conditional random field (CRF), 1, 8, 10,
   17, 31, 125, 151
   loss, 127, 129, 131, 133
   model, 10, 31, 127, 131, 136

Printed in the United States
by Baker & Taylor Publisher Services